# Advances in Experimental Medicine and Biology

## Volume 1465

*Advances in Experimental Medicine and Biology* provides a platform for scientific contributions in the main disciplines of the biomedicine and the life sciences. This series publishes thematic volumes on contemporary research in the areas of microbiology, immunology, neurosciences, biochemistry, biomedical engineering, genetics, physiology, and cancer research. Covering emerging topics and techniques in basic and clinical science, it brings together clinicians and researchers from various fields.

*Advances in Experimental Medicine and Biology* has been publishing exceptional works in the field for over 40 years, and is indexed in SCOPUS, Medline (PubMed), EMBASE, Reaxys, EMBiology, the Chemical Abstracts Service (CAS), and Pathway Studio.

2023 CiteScore: 5.9

Sanchita Bhatnagar
Editor

# Epigenetic Mechanisms in Breast Cancer Therapy and Resistance

Springer

*Editor*
Sanchita Bhatnagar
Medical Microbiology and Immunology
UC Davis Health System
Sacramento, CA, USA

ISSN 0065-2598 ISSN 2214-8019 (electronic)
Advances in Experimental Medicine and Biology
ISBN 978-3-031-66685-8 ISBN 978-3-031-66686-5 (eBook)
https://doi.org/10.1007/978-3-031-66686-5



This Springer imprint is published by the registered company Springer Nature Switzerland AG
The registered company address is: Gewerbestrasse 11, 6330 Cham, Switzerland

# Contents

# MDIG in Breast Cancer Progression and Metastasis

1

Chitra Thakur, Yiran Qiu, Zhuoyue Bi, Ziwei Wang, and Fei Chen

**Abstract**

Breast cancer is the most diagnosed cancer among women worldwide, and metastasis remains the major cause of breast cancer-related mortality and is associated with poor patient outcomes. Among breast cancers, triple-negative breast cancers have the worst prognosis owing to their highly aggressive and metastasizing attributes and hence have limited therapeutic options. Here, we have presented our research on an environmentally regulated gene named mdig and its role in the pathogenesis of breast cancer and metastasis. Through global proteomics, chromatin immunoprecipitation sequencing, and a mouse model of orthotopic xenograft, our studies established mdig as an anti-metastasis modulator in breast cancer with its influence on the methylation of DNA and histone proteins, thereby regulating the expression of genes implicated in epithelial-mesenchymal transitional, cell motility, and metastasis.

C. Thakur (✉) · Y. Qiu · Z. Bi · Z. Wang · F. Chen (✉)
Stony Brook Cancer Center and Department of Pathology, Renaissance School of Medicine, Stony Brook University, Stony Brook, NY, USA
e-mail: chitra.thakur@stonybrookmedicine.edu; Fei.Chen@stonybrookmedicine.edu

**Keywords**

Triple-negative breast cancer · Epigenetics · H3K36me3 · Metastasis · Invasion

## Abbreviations

| | |
|---|---|
| DCIS | ductal carcinoma in situ |
| IDC | invasive ductal carcinoma |
| ILC | invasive lobular carcinoma |
| HER2 | human epidermal growth factor receptor 2 |
| ER | estrogen receptor |
| PR | progesterone receptor |
| TNBC | triple-negative breast cancers |
| DDRT-PCR | differentiation display RT-PCR |
| *JmjC* | Jumonji C |
| OS | overall survival |
| DMFS | distant metastasis-free survival |
| RFS | relapse-free survival |
| PPS | post-progression survival |
| UPA | urokinase plasminogen activator |
| PAI-1 | inhibitor plasminogen activator inhibitor-1 |
| EMT | epithelial-mesenchymal transition |
| WT | wild-type |
| KO | knockout |
| ChIP-seq | chromatin immunoprecipitation sequencing |



S. Bhatnagar (ed.), *Epigenetic Mechanisms in Breast Cancer Therapy and Resistance*, Advances in Experimental Medicine and Biology 1465, https://doi.org/10.1007/978-3-031-66686-5_1

| | |
|---|---|
| ECM | extracellular matrix |
| 5hmC | 5-hydroxymethylcytosine |
| 5mC | 5-methylcytosine |

## 1.1 Breast Cancer

Breast cancer is the most diagnosed cancer worldwide, and it has surpassed even lung cancer. In fact, it has been estimated that in the current year 2023, 297,790 women in the United States will be diagnosed with invasive breast cancer and 55,720 women with noninvasive (in situ) breast cancer [1]. Interestingly, in the United States alone, one in eight women is predicted to have an appearance of breast cancer at some time point in their lives. Hence, breast cancer is an emerging concern with a huge burden on public health, and research toward the understanding of breast cancer development and progression is pivotal in designing better breast cancer therapies.

Breast cancer, being a highly heterogeneous disease, poses a significant challenge in fully delineating its origin and the unique underlying pathways orchestrating the development and metastasis of breast cancer. The anatomy of breast tissue comprises glands and connective tissues. While glands are milk-producing entities called lobules, connective tissues consist of fatty and fibrous tissues. The ducts fetch the milk to the nipples. Breast cancer originates in the ducts or the lobules and can either be benign or invasive based on the extent of its spread, called metastasis. Hence, a noninvasive, early-stage breast cancer limited to the ducts is called ductal carcinoma in situ (DCIS), whereas invasive cancer that originated in the ducts or lobules is referred to as invasive ductal carcinoma (IDC) and invasive lobular carcinoma (ILC), respectively [1, 2].

Based on the expression of hormone receptors such as the human epidermal growth factor receptor 2 (HER2), estrogen receptor (ER), progesterone receptor (PR), proliferation statuses of the tumors as assessed by Ki67 marker and/or mutations in the "Breast Cancer BRCA" gene, molecular characterization of breast cancer has identified its distinct subtypes such as the Luminal A, Luminal B, Her2-enriched, and basal/triple-negative breast cancers. These different subtypes of breast cancer have been associated with specific clinical responses and hence have crucial relevance in breast cancer therapies [3, 4].

Almost 10% of all breast cancers have been attributed to family history or genetic predisposition and mutations in the BRCA gene; *BRCA1* and *BRAC2* represent the most common germline alterations linked with breast cancer, possessing a cumulative 70% lifetime threat regarding the appearance of breast cancer [5, 6]. Studies identified certain genes such as p53, Myc, PTEN, PIK3CA, ERBB2, CCND1, GATA3, and *FGFR* as the most recurring genes amplified in early breast tumors [7]. Moreover, patients bearing mutations in the *BRCA1*, *BRCA2*, *TP53*, and *PTEN* genes have an elevated risk of breast cancer development [8].

Metastasis remains the most significant cause of cancer-related mortality in breast cancer patients. Among the different subtypes of breast cancer, triple-negative breast cancers (TNBC) are highly aggressive due to their increased metastatic features and, hence, have the worst prognosis [9–11]. TNBC accounts for 15–20% of breast cancers and often metastasizes to the lungs, liver, brain, and bones [12].

## 1.2 External Factors in the Regulation of Breast Cancer

There are multiple determinants that influence a woman's likelihood of developing breast cancer. While breast cancer etiology is complex, there isn't a single cause, but numerous, that can be listed under modifiable and non-modifiable risk factors [13]. Lifestyle aspects such as obesity, lack of physical activity, alcohol consumption, smoking, and exposure to carcinogens are the modifiable factors, whereas reproductive and hormonal risk factors like female gender, old age, breast tissue density, family history of breast or ovarian cancer, and germline mutations of high-penetrance genes such as *BRCA1* and *BRCA2* constitute some of the non-modifiable risk factors [9]. In addition to this, prolonged exposure to estrogen is yet another risk factor, where early menarche, delayed menopause, not breastfeeding, etc., may elevate the woman's lifelong exposure to estrogen. Interestingly, there exists an inverse relationship between physical activity and breast cancer incidence [14], suggesting that obesity or lack of exercise should be considered as one of the important modifiable factors that should be addressed by implementing physical activities among the population [15].

Other contributing factors include environmental exposure to hazardous agents via the air, food, or water, the humans consume, and the products they utilize. In this context, research suggests the impact of environmental factors and living surroundings in high-income countries where there has been an increased breast cancer incidence [16]. Industrial and agricultural chemicals such as benzene, ethylene oxide, 1,3-butadiene, polychlorinated biphenyls, pesticides, polycyclic aromatic hydrocarbons, and dioxin and endocrine disruptors such as bisphenol A, parabens, phthalates, polybrominated diphenyl ethers, perfluoroalkyl substances, phytoestrogens, ionizing radiations, tobacco, and metals like lead and cadmium are some of the environmental agents implicated in carcinogenesis and are well described [13]. In fact, almost 216 chemicals have been identified that are able to generate breast tumors in animal models, and strikingly, among those chemicals identified, 50% are the ones that women are commonly exposed to on a daily basis [17–19].

Such chemicals can make the breast tissue vulnerable to carcinogenesis and alter normal tissue homeostasis. Moreover, they can mimic certain hormones and can cause DNA damage, altogether serving as a great recipe for breast carcinogenesis. Notably, studies have indicated that the timing of exposure really makes a difference, where early life exposures increase the chances of breast cancer development later on in life [20].

To establish the kinds of risks and exposures that contribute to breast cancer development, one must investigate the gene-environment interaction aspect to further advance the knowledge of the genetic repertoire of the affected tissue because of environmental exposures. Moreover, it is pivotal to understand the genetic and epigenetic mechanisms orchestrated in breast cancer development under occupational and environmental settings.

## 1.3 Mdig, an Environmental Regulated Gene

**M**ineral **d**ust-**i**nduced **g**ene (mdig; also known as RIOX2, NO52, or MINA53) is an environmentally regulated gene and has critical implications in human cancers. Historically, this gene was discovered in the mid-1990s during a study where researchers were exploring the inflammatory cytokines in occupational lung diseases. Mdig was identified via differentiation display RT-PCR (DDRT-PCR) conducted on mRNAs isolated from the alveolar macrophages of the population with chronic lung diseases resulting from exposures to mineral dust in the mining occupation/industry field [21].

After its initial discovery, we, including several other groups, reported its existence and its role in human neoplasia. For instance, in 2002, a specific gene called the myc-induced nuclear antigen, MINA53, was identified in the human glioblastoma cell line T98G cells with overexpressed myc. Interestingly, MINA53 was found to be in 100% homology with mdig that was originally found in the alveolar macrophages [22]. Subsequently, mdig was also identified as a nucleolar protein in an amphibian *Xenopus laevis* and was termed as NO52 by the researchers [23].

The human mdig gene resides on chromosome 3 (3q12.1) and contains 12 exons spanning a 30 kb region with a coding sequence comprising 1398 bp that encodes a protein consisting of 465 amino acids with a molecular weight of 53 kDa. Predominately, mdig is a nuclear protein; however, the nucleolus also harbors a part of the protein. One of the unique features of mdig is the presence of a conserved Jumonji C (*JmjC*) domain in the mdig protein. Also, several studies highlighted that the JmjC family proteins exhibit histone demethylase activity [24–26].

Interestingly, because JmjC domain has been recognized as a signature motif of the JmjC family of histone demethylases [25], it is apt to expect that mdig would also be involved in the histone modifications, hence influencing the epigenetic mechanisms [27]. In fact, histone demethylase-like activity was indeed detected in the cells with mdig overexpression or knockdown [28]. In human lung cancer samples, an inverse link between mdig and H3K9me3 was observed [28]. Also, a moderate demethylase activity was found in a test tube reaction of the H3K9me3 peptide and the immunoprecipitated mdig protein [29]. Our most recent study revealed mdig's regulation on the three major repressive histone trimethylation markers H3K9me3, H3K27me3, and H4K20me3 in a human bronchial epithelial cell line (BEAS-2B cells, A549 cells) and triple-negative breast cancer cell line (MDA-MB-231 cells) [30]. Through proteomics analysis, we have identified the interaction of mdig with many other proteins belonging to the chromatin binding proteins and proteins involved in DNA repair, replication, and RNA splicing [30].

Mdig has oncogenic roles and is involved in epigenetic regulation in human lung and breast cancers, where mdig's expression was found to be increased [27, 31]. This also stems from the fact that mdig expression is under the direct control of oncoprotein c-myc [22] and that mdig exhibits pro-proliferative functions [32, 33] where its silencing delays the cell-cycle transition from G1 to S phase [21]. Some studies suggested the involvement of mTOR complex 2/Akt and pyruvate dehydrogenase kinase 1/Akt signaling pathways in mdig-mediated cell proliferation [34]. Apart from c-myc, certain environmental agents such as mineral dust, tobacco smoke, arsenic, and silica can induce the expression of mdig [21, 35–37], where JNK-STAT3 signaling was identified as a possible mechanism behind this process [36]. Most recently, mdig's role in liver regeneration via regulation of histone methylation was reported. Mdig exerted the demethylation of H3K9me3, further upregulating myc by the activation of OTX2, ultimately assisting the regeneration of liver [38]. Mdig also has crucial roles in immune regulation where it is implicated in allergen-induced inflammation and is involved in the differentiation of proinflammatory cytokine Th 17 cells [39–41].

Mdig expression is found to be elevated in certain human cancers such as in breast cancer [42], colon cancer [43], esophageal squamous cell carcinoma [44], gingival squamous cell carcinoma [45], lymphoma [46], renal cell carcinoma [47], glioblastoma [48], gastric cancer [49], hepatocellular carcinoma [50], cholangiocarcinoma [51], multiple myeloma [35], pancreatic cancer [52], etc. In certain cancers, mdig levels correlated with the patients overall survival, where high mdig expression was associated with poor overall survival in squamous lung cancer and breast cancer without lymph node positivity [36, 42]. In prostate cancer, mdig upregulation was firmly linked with the tumor stage, TP53 mutation, Gleason score, and disease-specific survival [53]. In renal cell carcinoma, mdig is suggestive of cancer progression and metastasis spread [54].

Our work also indicated that mdig regulates the expression of genes in cell growth and stemness of normal and cancer stem cells [55]. Moreover, in lung adenocarcinoma, mdig facilitated cisplatin resistance, which was mediated by mdig's regulation of the ABC transporter through the activation of Wnt/β-catenin signaling [34].

## 1.4 Mdig in Breast Cancer and Metastasis

Our studies revealed that mdig expression levels impact the prognosis of lung and breast cancer. In breast cancer, an inverse correlation between mdig expression and patient survival, such as poorer overall survival (OS), distant metastasis-free survival (DMFS), relapse-free survival (RFS), and post-progression survival (PPS), was identified. In fact, elevated mdig levels predicted poorer overall survival in Luminal A breast cancer patients. In contrast, breast cancer patients with lymph node metastasis exhibited better overall survival with increased expression of mdig (Figs. 1.1a, and 1.1b) [42].

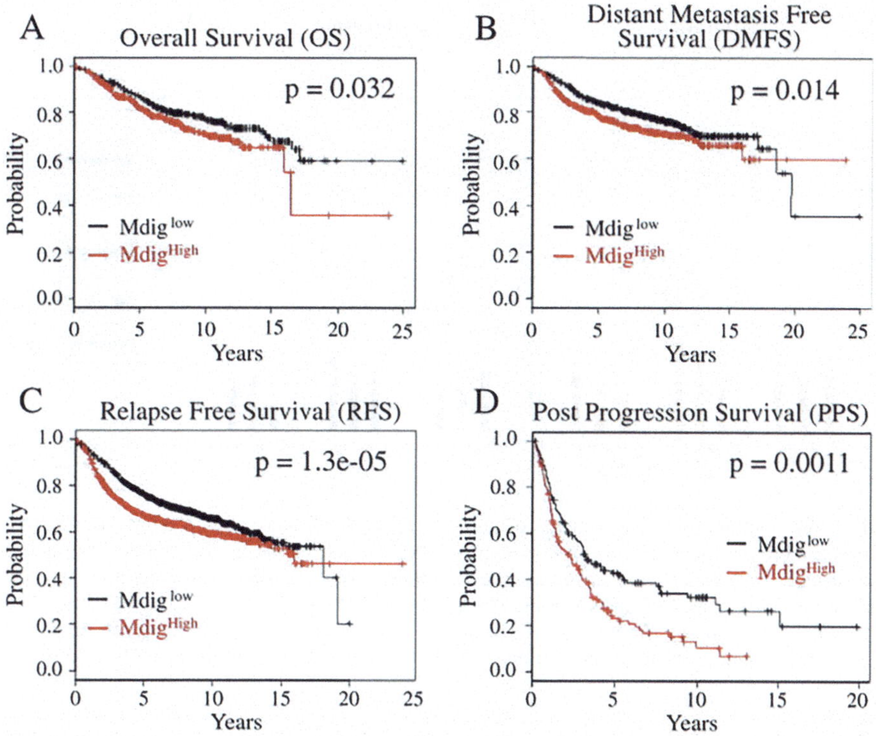


**Fig. 1.1a** Increased expression of mdig predicates poorer overall survival (OS, **a**), distant metastasis-free survival (DMFS, **b**), relapse-free survival (RFS, **c**), and post-progression survival (PPS, D) of the breast cancer patients [42]

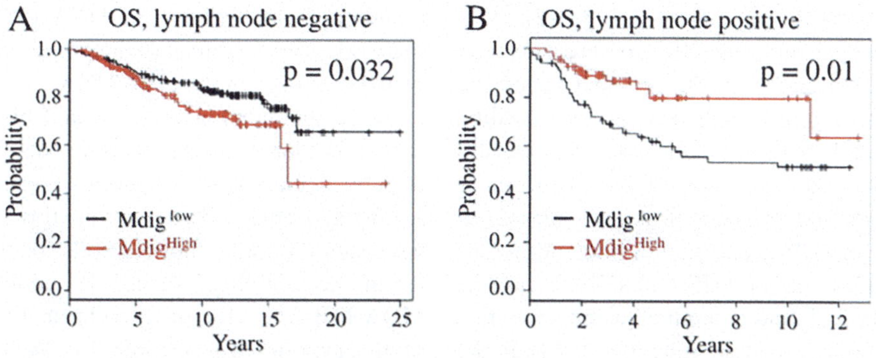


**Fig. 1.1b** The lymph node status determines the predictive power of mdig expression on patient survival. (**a**). Increased expression of mdig predicates poorer OS of the breast cancer patients who were lymph node negative. (**b**). Increased expression of mdig predicates better OS of the breast cancer patients who were lymph node positive [42]

Intrigued by these initial findings of mdig's influence in predicting patient survival, we dwelled deeper to understand the role mdig plays in breast tumorigenesis. Using breast cancer cell lines, mdig silencing approach, and immunohistochemical evaluation of mdig expression in early and late-stage breast cancer patient tissues, we showed the differential roles of mdig in early- and late-stage breast cancer. Proliferation assays conducted on the normal breast cells MCF10A and breast cancer cells MDA-MB-231, MCF-7, and T47-D with mdig silencing revealed that

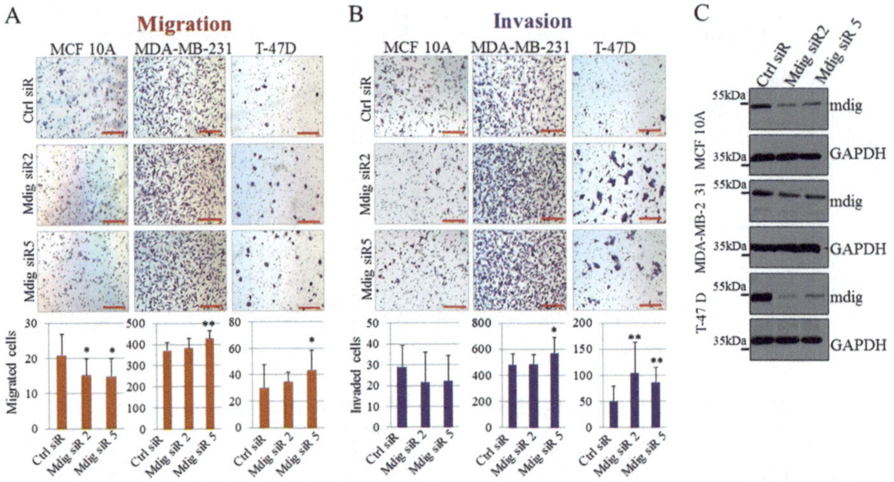


**Fig. 1.2** Mdig affects cell motility and invasion of noncancerous breast and breast cancer cells. Transwell migration (**a**) and Matrigel invasion (**b**) assay of indicated cell lines following transfection with Ctrl siRNA, mdig siR2, or mdig siR5, 24 h post transfection. Scale bar = 200 μM. The quantification of the number of migrated cells and invaded cells is depicted by a bar graph below each of the figure panels. Ten randomly selected fields were counted for migrated and invasive cells ($n = 10$). *$p < 0.05$; **$p < 0.01$. (**c**) Western blotting shows a decrease in mdig protein for indicated cell lines 24 h after knockdown of mdig by siRNAs [55]

mdig is pro-proliferative for normal breast cells. However, its silencing enhanced the migration and invasion potential of breast cancer cells. This suggests mdig's negative regulation on breast cancer cell motility and invasion (Fig. 1.2), [55].

Since mdig exerted its effect on breast cancer cell migration and invasion, we next tested the expression of genes implicated in cell motility, invasion, and metastasis. Interestingly, we did find an elevated expression of the extracellular matrix-degrading protease urokinase plasminogen activator (UPA) and its inhibitor plasminogen activator inhibitor-1 (PAI-1), CXCR4, and its ligand CXCL12, and an epithelial-mesenchymal transition (EMT) marker, vimentin (VIM), in the mdig-silenced MDA-MB-231 cells when compared to the control counterparts [55].

As mdig has a conserved JmjC domain with some demethylase functions, we evaluated the methylation of DNA and histone H3 lysine 9 protein in normal breast cells and breast cancer cells. Measuring the global DNA methylation by assessing 5-methylcytosine (5mC) and 5-hydroxymethylcytosine (5hmC) in cells transfected with mdig siRNAs showed a remarkable increase in 5mC in MCF10A, MDA-MB-231, and T-47D cells silenced for mdig. This observation suggested that loss of mdig favored DNA methylation, further attesting the fact that mdig serves as a DNA demethylase or a cofactor that facilitates demethylation of DNA [55]. Hence, mdig-silenced cells lead to the piling up of methylated cytosines in the cell's DNA.

Since methylation of DNA and histone proteins modulates the accessibility of chromatin, thereby mediating the formation of open (euchromatin) and closed (heterochromatin) structures, we tested the mdig-silenced cells for their chromatin accessibility status. In mdig-silenced MDA-MB-231 cells, an open chromatin structure of the genome was favored for the metastatic genes CXCL12, CXCR4, MMP-1, and MMP-9. These findings clearly demonstrated the influence of mdig on the DNA and histone methylation, as well as on the chromatin accessibility for the genes implicated in cell motility and invasion [55].

Immunohistochemical evaluation of mdig protein on breast cancer tissue microarray revealed a decreased expression of mdig in the late-stage lymph node metastasis, further suggesting that loss of mdig is an important feature in metastasis and mdig's inhibitory role on the invasion and metastasis of breast cancer cells was recognized [55, 56]. A similar trend was observed in lung cancer where we identified mdig's loss of expression in increasing tumor grade and stage, whereas the early-stage cancer exhibited strong mdig expression [57]. These findings suggest that mdig's role in human lung and breast cancer may be context dependent with differential mdig expression in the early vs. late stage and metastatic vs. non-metastatic scenarios.

Therefore, it is crucial to understand how mdig affects the invasion potential of breast cancer cells. To get a better understanding of the underlying mechanism, we utilized the CRISPR-Cas9 gene editing approach for mdig silencing in MDA-MB-231 cells in conjunction with quantitative proteomics, global ChIP-seq analysis, and an orthotopic mouse breast cancer model.

Loss of mdig is a characteristic feature of aggressive breast cancers. Comparing the expression status of mdig protein via immunohistochemistry in hormone-positive vs. TNBC patient tissues showed that mdig expression is reduced significantly in TNBC cases. This finding was further supported by the fact that patients with lymph node-negative metastasis have a shorter overall survival in comparison to those patients who were positive for lymph node metastasis as evaluated by Kaplan-Meier survival curves of large breast cancer patient's dataset. These findings prompted us to expand our studies in animal models of experimental tumorigenesis and metastasis to further decipher the role of mdig in breast cancer progression.

For that, we engineered triple-negative breast cancer cells with stable knockdown for mdig via the CRISPR Casa 9 gene editing approach. Wild-type (WT) MDA-MB-231 cells and mdig knockout (KO) MDA-MB-231 cells were injected into the mammary fat pads of the severely immunocompromised NSG mice, and tumors were measured after 7 weeks' time. Compared to the WT tumors, the tumors from the KO mice were remarkably smaller in size (Fig. 1.3) [58]. This suggests that mdig is important for breast cancer formation and that mdig deficiency was sufficient to impede the primary tumor formation in vivo. This suggest that mdig has crucial roles to play in tumor cell proliferation and growth during the initial stages of neoplasia establishment. Interestingly, evaluating the livers of the NSG mice, mdig KO mice showed enhanced micrometastasis in the liver as assessed by cytokeratin 18 staining positivity of the liver lesions [58].

These animal data corroborated our in vitro findings where mdig exerted its pro-proliferative effect and anti-metastasis influence on breast cancer cells. To gain an insight into the underlying features of mdig where mdig deficiency increased

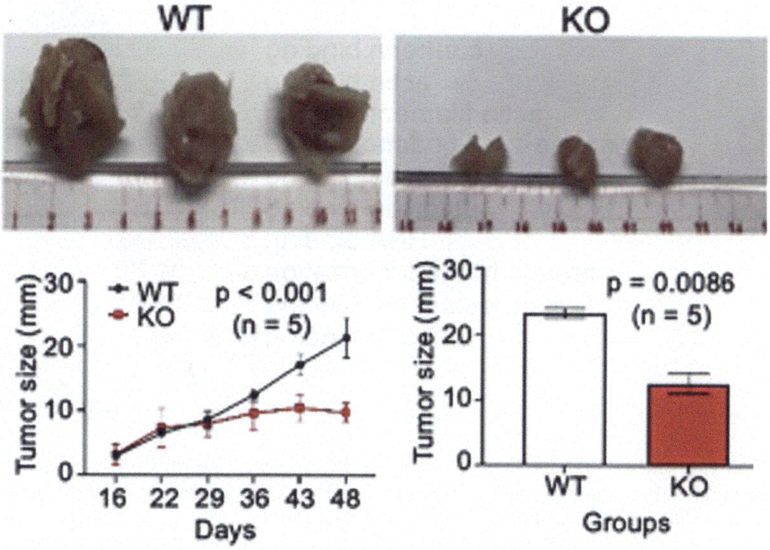


**Fig. 1.3** Breast tumor formation in NSG mice injected with WT and mdig KO MDA-MB-231 cells. Photographs of the freshly resected mammary tumors from the mice injected with WT and KO cells, respectively. The WT cells form larger tumors than the KO cells. Lower panels show quantification of the tumor size. Data are represented as mean ± SD [58]

the metastatic propensity of breast cancer cells, we subjected the WT and KO cells to global proteomics and chromatin immunoprecipitation sequencing (ChIP-seq) analysis. Analyzing the raw data for the determination of fold change of proteins as a normalized ratio for KO cells in comparison to the WT cells showed an enrichment of genes related to cell motility and metastasis as the topmost upregulated genes identified in the KO cells. A distinct proteomic profile was observed between the WT and KO cells, and gene ontology analysis of the molecular functions revealed that in the KO cells, the downregulated proteins are the ones implicated in rRNA processing, rRNA metabolism, fatty acid oxidation, and biogenesis of ribosome. This finding is in accordance with the concept of mdig's localization in nucleoli linked with ribosomal functions. In addition, proteins that were upregulated in the KO cells belonged to the pathways associated with binding to cadherin, actin, ubiquitin, GTPase, etc., indicating that proteins bestowing cells with motility and invasion were increased in the KO cells (Fig. 1.4) [58]. This observation was further strengthened by the migration-invasion assay, where the KO cells exhibited a substantial enrichment of invasive migration when compared to the WT counterparts [58].

Interestingly, proteins that showed a twofold increase in the proteomics analysis in the KO cells relative to WT, such as MAGED2, CTSD, and STMN1, have roles in cancer metastasis.

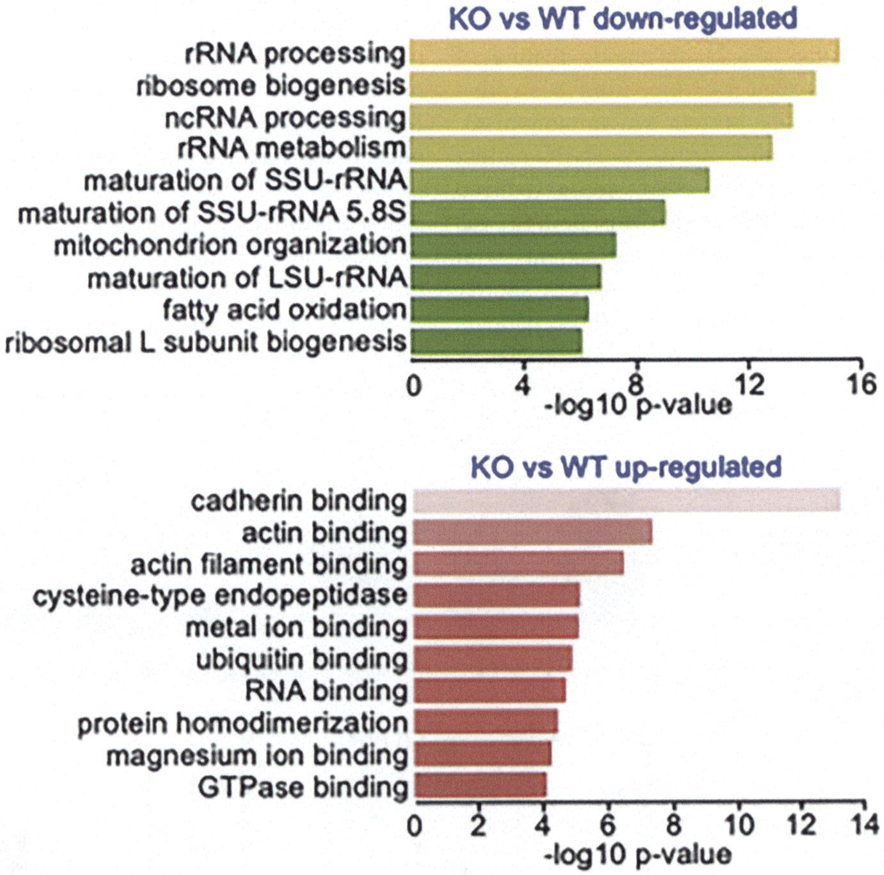


**Fig. 1.4** Quantitative proteomics analysis for protein expression between WT and KO MDA-MB-231 cells, showing the ontology GO molecular function assay of the downregulated proteins (upper panel) and the upregulated proteins (bottom panel) in the KO cells [58]

Notably, ChIP-seq data from the WT and KO cells showed a substantial enrichment of the repressive histone trimethylation markers, H3K9me3 and H4K20me3, in the KO cell category. Surprisingly, an active transcriptional elongation marker H3K36me3 was significantly amplified on the gene bodies in the KO cells relative to WT cells, further suggesting that mdig is antagonistic toward H3K9me3.

In addition to the repressive histone trimethylation markers, we also checked the status of H3K36me3, an active transcriptional elongation marker of the genes, in both WT and mdig KO cells. A pronounced magnification of H3K36me3 on the gene bodies in the KO cells relative to the WT cells was noted, indicating that mdig is also antagonistic toward H3K9me3, H4K20me3, and H3K36me3 cells. Moreover, around 566 genes with profound enrichment of H3K36me3 and 184 genes with a reduced enrichment of H3K36me3 were identified in the KO cells. To ascertain the molecular function of these identified genes with increased H3K36me3, we conducted pathway analysis, which revealed that the top-ranked pathways are associated with β1 integrin interaction, DNA base excision repair, and Gα signaling.

As critical regulators of metastasis and angiogenesis, integrins are well-recognized cell surface receptors of several extracellular matrix (ECM) proteins [59–61]. Furthermore, in breast cancer progression and metastasis, β1 integrin has crucial implications [60, 62]. Since histone markers preside over a myriad of cellular processes, our findings shed light on the fact that in breast cancer, mdig disruption enriches the H3K36me3 marker that, in turn, exerts a positive regulation on the expression of genes involved in DNA damage repair, angiogenesis, and metastasis. Among the topmost 25 upregulated proteins in the KO cells that exhibited elevated enrichment of H3K36me3 are the ones belonging to the family of proteins, namely, IQSEC2, HUWE1, MAGED2, UPF3B, LAMP2, CUL4B, etc. These genes and others have significant implications in angiogenesis, EMT, glycolysis, stemness, and cancer metastasis, for example, the IQSEC2 and its family members, i.e., the Arf-nucleotide exchange factors [63], NMNAT1 [64], MAGED2 [65], SLC12A7 [66], IGFBP7 [67], HUWE1 [68], PLS3 [69], PRNP [70], HMGB2 [71], LAMP2 [72], and CTSD [73]. These findings showed a strong correlation between the H3K36me3 enrichment status, and the major genes involved in breast cancer progression and metastasis processes, where elevated H3K36me3 would preferentially favor the expression of those genes.

H3K36me3 has been linked with elongating Pol II and is enriched within the body of transcriptionally active genes. Our data indicate that elevated H3K36me3 is pivotal for the expression of the identified top 25 ranked genes that are most likely to contribute to the metastasis feature in the KO cells. For example, proteins regulating the cytoskeleton assembly, with frequent mutations in human lung adenocarcinoma, is IQSEC2 [74]. Also, family members of IQSEC proteins, such as the IQSEC1, are associated with the MDA-MB-231 breast cancer cells metastasis [75]. Likewise, NMNAT1 has been shown in tumors associated with high grades, metastasis, and poor survival [76]. Also, among the top 25 identified upregulated proteins are the lysosomal proteins PRNP, LAMP2, NPC2, and CTSD. Lysosomal proteases are implicated in the remodeling of the extracellular matrix and hence have important roles in breast cancer invasion [77]. LAMP2 and CTSD have been recognized as important propellers of metastasis in breast cancer cells [72, 73].

One of the topmost genes that was identified in our ChIP-seq analysis and proteomics was MAGED2, which was presented with a more than sevenfold increase in its expression along with an enriched H3K36me3 in the KO cells. Therefore, we focused our attention on MAGED2 and silenced its expression in the KO cells by using the siRNA approach. Then, we conducted the migration-invasion assay to test the invasive potential of KO cells with or without MAGED2. Interestingly, the knockdown of MAGED2 showed a substantial abrogation of the invasive capabilities of the KO cells. Next, we tested MAGED2's clinical relevance and found that in TNBC specifically, increased MAGED2 expression serves as a poor prognostic factor for the overall survival of patients. In addition to this,

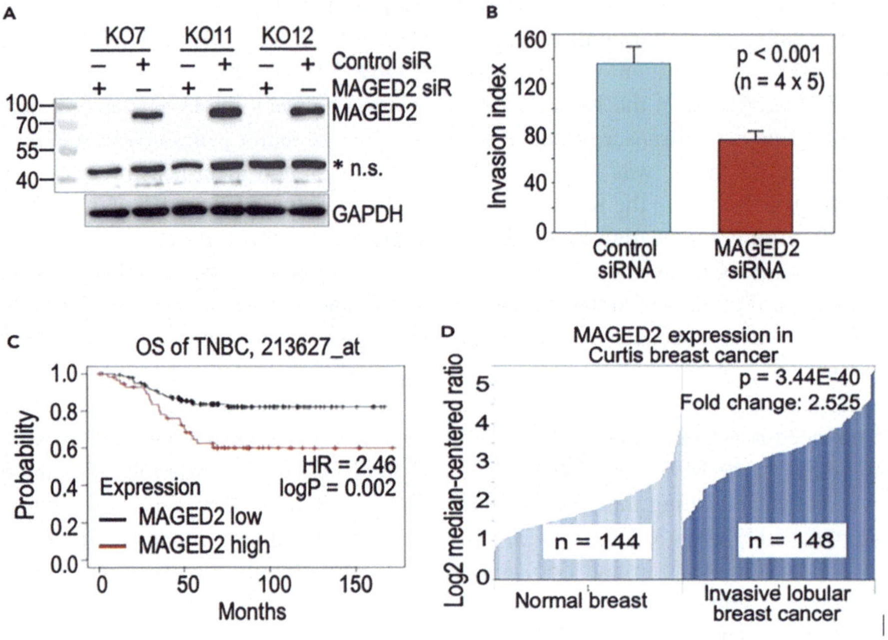


**Fig. 1.5** MAGED2 positively regulates the invasive migration of breast cancer cells. (**a**) Three clones of mdig KO MDA-MB-231 cells were silenced for the MAGED2 gene using siRNA targeting MAGED2. GAPDH was used as loading control for Western blotting. MAGED2 was completely silenced by the MAGED2 siRNA but not the control siRNA. (**b**) Loss of MAGED2 through siRNA silencing resulted in lower invasion capability of the KO cells, indicating MAGED2 is a positive regulator for invasive migration. Data are represented as mean ± SD. (**c**) High MAGED2 predicts poorer relapse-free survival (RFS) of breast cancer patients concomitant with p53 mutation. (**d**) High MAGED2 expression is a common feature of invasive breast cancer. Data are derived from Curtis breast cancer that showed a more than twofold elevation of MAGED2 gene in invasive lobular breast cancer compared to normal breast tissue [58]

our studies identified that increased MAGED2 is a common feature of invasive breast cancers, where its expression is in conjunction with other genes critical for bestowing malignant phenotypes in breast cancer (Fig. 1.5) [58]. Moreover, MAGED2 and other genes implicated in EMT, and metastasis negatively correlated with mdig. Altogether, our studies demonstrate that MAGED2 is an essential regulator of breast cancer cell motility and that it positively regulates the metastasis of breast cancer.

## 1.5 Conclusion

Our findings provided the first evidence demonstrating the anti-metastasis role of mdig in breast cancer. Since a decreased expression of mdig has been observed in late-stage breast cancer [55] and lung cancer [57], our results suggest differential roles of mdig during the early and late stages of tumorigenesis. Mdig fosters the expression of oncogenes like c-Myc, EGFR, and others owing to its antagonistic properties on the repressive histone trimethylation markers such as H3K9me3, H3K27me3, and H4K20me3. Hence, with enhanced expression of c-Myc and EGFR, tumor initiation and growth are promoted during the early phases of cancer development [78]. However, in the late stages, when the tumors are well developed, there's an amplified enrichment of H3K36me3 owing to mdig's dwindled expression in the tumors. Such an elevated H3K36me3 boosts the transcription and expression of genes implicated in EMT and metastasis (Fig. 1.6) [58].

As mdig has been strongly associated with various human cancers, it would be important to understand its biological roles where it can be tar-

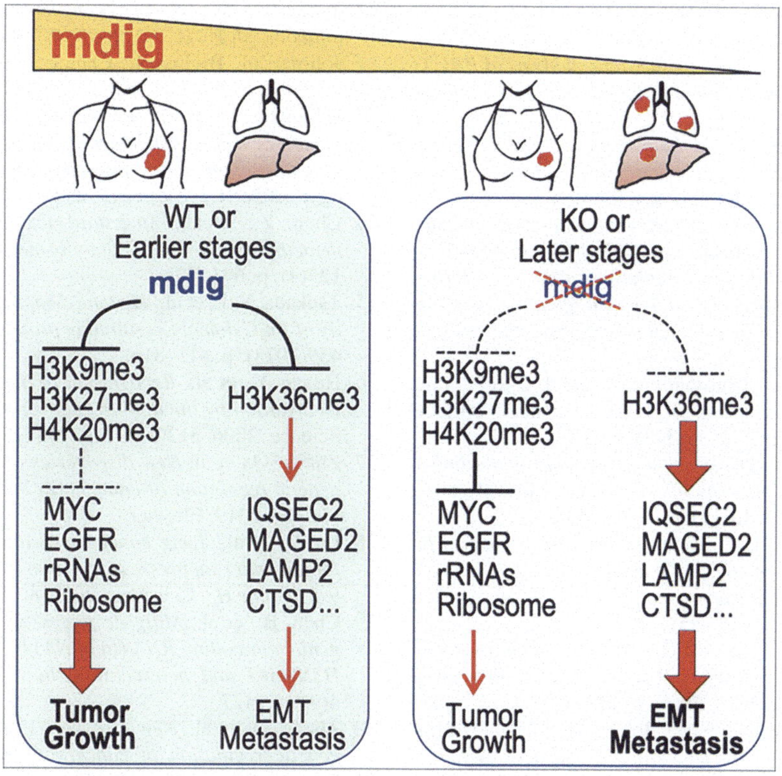


**Fig. 1.6** Schematic diagram shows the regulatory roles of mdig on histone protein methylations, tumor growth, and metastasis. In earlier-stage breast cancer or WT breast cancer cells, increased expression of mdig antagonizes the repressive trimethylation of histone proteins, including H3K9me3, H3K27me3, and H4K20me3, leading to derepression of the major oncogenes, such as MYC, EGFR, and others, that promote tumor growth. Mdig is also antagonistic toward H3K36me3 and preferentially enforces the expression of the genes in cancer cell motility, EMT, and metastasis. In later-stage breast cancer or the mdig KO breast cancer cells, because of the loss of mdig expression, the repressive trimethylations of histone proteins compromise the expression of the growth-promoting oncogenes. Meanwhile, loss of mdig causes enhanced enrichment of H3K36me3 that boosts transcriptional elongation of the genes involved in cell motility, EMT, and metastasis. Thus, metastasis or uplifted metastatic potential is dominant in the later-stage breast cancer, or the breast cancer cells with mdig knockout [58]

geted for cancer therapies. Most recently, investigators designed an inhibitor of MINA53 and evaluated its anticancer properties in terms of proliferation, substrate binding, cellular selectivity, and sensitization in solid cancer cell lines. Their results show antiproliferative action and cancer cells' responsiveness to conventional chemotherapy, highlighting the importance of such research on mdig inhibitors in oncology settings [79].

Therefore, studying the functional dynamics and expression of mdig, methylation of DNA and histones, and the key players of EMT and metastasis is indispensable for designing new strategies for targeted therapies in breast cancer and other malignancies.

## References


1. Henry, N.L. and L.A. Cannon-Albright, *Breast cancer histologic subtypes show excess familial clustering.* Cancer, 2019. **125**(18): p. 3131–3138.
2. Tan, P.H., et al., *The 2019 World Health Organization classification of tumours of the breast.* Histopathology, 2020. **77**(2): p. 181–185.
3. Zardavas, D., et al., *Clinical management of breast cancer heterogeneity.* Nat Rev Clin Oncol, 2015. **12**(7): p. 381–94.
4. Malhotra, G.K., et al., *Histological, molecular and functional subtypes of breast cancers.* Cancer Biol Ther, 2010. **10**(10): p. 955–60.
5. Kuchenbaecker, K.B., et al., *Risks of Breast, Ovarian, and Contralateral Breast Cancer for BRCA1 and*

*BRCA2 Mutation Carriers.* Jama, 2017. **317**(23): p. 2402–2416.
6. Chen, S. and G. Parmigiani, *Meta-analysis of BRCA1 and BRCA2 penetrance.* J Clin Oncol, 2007. **25**(11): p. 1329–33.
7. Nik-Zainal, S., et al., *Landscape of somatic mutations in 560 breast cancer whole-genome sequences.* Nature, 2016. **534**(7605): p. 47–54.
8. Walsh, T. and M.C. King, *Ten genes for inherited breast cancer.* Cancer Cell, 2007. **11**(2): p. 103–5.
9. Fragomeni, S.M., A. Sciallis, and J.S. Jeruss, *Molecular Subtypes and Local-Regional Control of Breast Cancer.* Surg Oncol Clin N Am, 2018. **27**(1): p. 95–120.
10. Jones, R.L., A. Constantinidou, and J.S. Reis-Filho, *Molecular classification of breast cancer.* Surgical pathology clinics, 2012. **5**(3): p. 701–717.
11. Dent, R., et al., *Triple-negative breast cancer: clinical features and patterns of recurrence.* Clinical cancer research, 2007. **13**(15): p. 4429–4434.
12. André, F. and C.C. Zielinski, *Optimal strategies for the treatment of metastatic triple-negative breast cancer with currently approved agents.* Ann Oncol, 2012. **23 Suppl 6**: p. vi46–vi51.
13. Hiatt, R.A. and J.G. Brody, *Environmental Determinants of Breast Cancer.* Annual Review of Public Health, 2018. **39**(1): p. 113–133.
14. Gammon, M.D., E.M. John, and J.A. Britton, *Recreational and Occupational Physical Activities and Risk of Breast Cancer.* JNCI: Journal of the National Cancer Institute, 1998. **90**(2): p. 100–117.
15. Conroy, S.M., et al., *Contextual Impact of Neighborhood Obesogenic Factors on Postmenopausal Breast Cancer: The Multiethnic Cohort.* Cancer Epidemiol Biomarkers Prev, 2017. **26**(4): p. 480–489.
16. Ferlay, J., et al., *GLOBOCAN 2008, cancer incidence and mortality worldwide: IARC CancerBase No. 10. Lyon: International Agency for Research on Cancer; 2010.* 2013.
17. Rudel, R.A., et al., *Chemicals causing mammary gland tumors in animals signal new directions for epidemiology, chemicals testing, and risk assessment for breast cancer prevention.* Cancer, 2007. **109**(S12): p. 2635–2666.
18. Brody, J.G., et al., *Environmental pollutants and breast cancer.* Cancer, 2007. **109**(S12): p. 2667–2711.
19. Rudel, R.A., et al., *New Exposure Biomarkers as Tools for Breast Cancer Epidemiology, Biomonitoring, and Prevention: A Systematic Approach Based on Animal Evidence.* Environmental Health Perspectives, 2014. **122**(9): p. 881–895.
20. Rodgers, K.M., et al., *Environmental chemicals and breast cancer: An updated review of epidemiological literature informed by biological mechanisms.* Environmental Research, 2018. **160**: p. 152–182.
21. Zhang, Y., et al., *The Human mineral dust-induced gene, mdig, is a cell growth regulating gene associated with lung cancer.* Oncogene, 2005. **24**(31): p. 4873–4882.
22. Tsuneoka, M., et al., *A novel myc target gene, mina53, that is involved in cell proliferation.* Journal of Biological Chemistry, 2002. **277**(38): p. 35450–35459.
23. Eilbracht, J., et al., *Protein NO52—a constitutive nucleolar component sharing high sequence homologies to protein NO66.* European journal of cell biology, 2005. **84**(2–3): p. 279–294.
24. Chen, Z., et al., *Structural insights into histone demethylation by JMJD2 family members.* Cell, 2006. **125**(4): p. 691–702.
25. Tsukada, Y.-I., et al., *Histone demethylation by a family of JmjC domain-containing proteins.* Nature, 2006. **439**(7078): p. 811–816.
26. Huang, Y., et al., *Recognition of histone H3 lysine-4 methylation by the double tudor domain of JMJD2A.* Science, 2006. **312**(5774): p. 748–751.
27. Zhang, Q., et al. *New discoveries of mdig in the epigenetic regulation of cancers.* in *Seminars in cancer biology*. 2019. Elsevier.
28. Lu, Y., et al., *Lung cancer-associated JmjC domain protein mdig suppresses formation of tri-methyl lysine 9 of histone H3.* Cell cycle, 2009. **8**(13): p. 2101–2109.
29. Chen, B., et al., *Mdig de-represses H19 large intergenic non-coding RNA (lincRNA) by down-regulating H3K9me3 and heterochromatin.* Oncotarget, 2013. **4**(9): p. 1427.
30. Zhang, Q., et al., *New discoveries of mdig in the epigenetic regulation of cancers.* Seminars in Cancer Biology, 2019. **57**: p. 27–35.
31. Thakur, C. and F. Chen, *Current understanding of mdig/MINA in human cancers.* Genes & Cancer, 2015. **6**(7–8): p. 288.
32. Xu, X., et al., *Effects of mdig on proliferation and apoptosis of lung cancer cells.* Oncol Lett, 2018. **16**(6): p. 7146–7151.
33. Yu, M., et al., *Paradoxical Roles of Mineral Dust Induced Gene on Cell Proliferation and Migration/Invasion.* PLOS ONE, 2014. **9**(2): p. e87998.
34. Geng, F., et al., *MDIG, a 2-oxoglutarate-dependent oxygenase, acts as an oncogene and predicts the prognosis of multiple types of cancer.* Int J Oncol, 2022. **61**(1).
35. Wu, K., et al., *Proteomic Characterization of the World Trade Center dust-activated mdig and c-myc signaling circuit linked to multiple myeloma.* Scientific reports, 2016. **6**(1): p. 36305.
36. Sun, J., et al., *Carcinogenic metalloid arsenic induces expression of mdig oncogene through JNK and STAT3 activation.* Cancer letters, 2014. **346**(2): p. 257–263.
37. Thakur, C., et al., *Oncoprotein mdig contributes to silica-induced pulmonary fibrosis by altering balance between Th17 and Treg T cells.* Oncotarget, 2015. **6**(6): p. 3722.
38. Du, J., et al., *MDIG-mediated H3K9me3 demethylation upregulates Myc by activating OTX2 and facilitates liver regeneration.* Signal Transduction and Targeted Therapy, 2023. **8**(1): p. 351.

39. Hemmers, S. and K.A. Mowen, *TH2 bias: Mina tips the balance.* Nature immunology, 2009. **10**(8): p. 806–808.
40. Yosef, N., et al., *Dynamic regulatory network controlling TH17 cell differentiation.* Nature, 2013. **496**(7446): p. 461–468.
41. Korn, T., et al., *IL-21 initiates an alternative pathway to induce proinflammatory TH17 cells.* Nature, 2007. **448**(7152): p. 484–487.
42. Thakur, C., et al., *Increased expression of mdig predicts poorer survival of the breast cancer patients.* Gene, 2014. **535**(2): p. 218–224.
43. Teye, K., et al., *Increased expression of a Myc target gene Mina53 in human colon cancer.* The American journal of pathology, 2004. **164**(1): p. 205–216.
44. Tsuneoka, M., et al., *Mina53 as a potential prognostic factor for esophageal squamous cell carcinoma.* Clinical cancer research, 2004. **10**(21): p. 7347–7356.
45. Kuratomi, K., et al., *Immunohistochemical expression of Mina53 and Ki67 proteins in human primary gingival squamous cell carcinoma.* The Kurume medical journal, 2006. **53**(3+ 4): p. 71–78.
46. Teye, K., et al., *Expression of Myc target gene mina53 in subtypes of human lymphoma.* Oncology reports, 2007. **18**(4): p. 841–848.
47. Ishizaki, H., et al., *Overexpression of the myc target gene Mina53 in advanced renal cell carcinoma.* Pathology international, 2007. **57**(10): p. 672–680.
48. Huang, M., et al., *MINA controls proliferation and tumorigenesis of glioblastoma by epigenetically regulating cyclins and CDKs via H3K9me3 demethylation.* Oncogene, 2017. **36**(3): p. 387–396.
49. Zhang, Q., et al., *Expression of Mina53 and its significance in gastric carcinoma.* The International journal of biological markers, 2008. **23**(2): p. 83–88.
50. Huo, Q., et al., *Dysfunction of IKZF1/MYC/MDIG axis contributes to liver cancer progression through regulating H3K9me3/p21 activity.* Cell Death & Disease, 2017. **8**(5): p. e2766–e2766.
51. Tan, X.-P., et al., *Upregulated expression of Mina53 in cholangiocarcinoma and its clinical significance.* Oncology letters, 2012. **3**(5): p. 1037–1041.
52. Kumar, S.A., et al., *Pathological and prognostic role of mdig in pancreatic cancer.* Genes & Cancer, 2017. **8**(7–8): p. 650.
53. He, C., et al., *Jumonji domain-containing protein RIOX2 is overexpressed and associated with worse survival outcomes in prostate cancers.* Frontiers in Oncology, 2023. **13**.
54. Ferreira, M.J., et al., *SETDB2 and RIOX2 are differentially expressed among renal cell tumor subtypes, associating with prognosis and metastization.* Epigenetics, 2017. **12**(12): p. 1057–1064.
55. Thakur, C., et al., *Loss of mdig expression enhances DNA and histone methylation and metastasis of aggressive breast cancer.* Signal Transduct Target Ther, 2018. **3**: p. 25.
56. Thakur, C., et al., *Loss of mdig expression enhances DNA and histone methylation and metastasis of aggressive breast cancer.* Signal Transduction and Targeted Therapy, 2018. **3**(1): p. 25.
57. Shi, J., et al., *Pathological and Prognostic Indications of the mdig Gene in Human Lung Cancer.* Cell Physiol Biochem, 2021. **55**(S2): p. 13–28.
58. Thakur, C., et al., *Deletion of mdig enhances H3K36me3 and metastatic potential of the triple negative breast cancer cells.* iScience, 2022. **25**(10): p. 105057.
59. Ganguly, K.K., et al., *Integrins and metastasis.* Cell Adh Migr, 2013. **7**(3): p. 251–61.
60. Gui, G.P., et al., *Integrin expression in primary breast cancer and its relation to axillary nodal status.* Surgery, 1995. **117**(1): p. 102–8.
61. Yousefi, H., et al., *Understanding the role of integrins in breast cancer invasion, metastasis, angiogenesis, and drug resistance.* Oncogene, 2021. **40**(6): p. 1043–1063.
62. Jonjić, N., et al., *Expression of beta-1 integrins on tumor cells of invasive ductal breast carcinoma.* Pathol Res Pract, 1993. **189**(9): p. 979–84.
63. D'Souza, R.S. and J.E. Casanova, *The BRAG/IQSec family of Arf GEFs.* Small GTPases, 2016. **7**(4): p. 257–264.
64. Yaku, K., et al., *NAD Metabolism in Cancer Therapeutics.* Front Oncol, 2018. **8**: p. 622.
65. Jia, B., et al., *Prognostic roles of MAGE family members in breast cancer based on KM-Plotter Data.* Oncol Lett, 2019. **18**(4): p. 3501–3516.
66. Brown, T.C., et al., *SLC12A7 alters adrenocortical carcinoma cell adhesion properties to promote an aggressive invasive behavior.* Cell Commun Signal, 2018. **16**(1): p. 27.
67. Paulitschke, V., et al., *Proteomic identification of a marker signature for MAPKi resistance in melanoma.* Embo j, 2019. **38**(15): p. e95874.
68. Vaughan, L., et al., *HUWE1 ubiquitylates and degrades the RAC activator TIAM1 promoting cell-cell adhesion disassembly, migration, and invasion.* Cell Rep, 2015. **10**(1): p. 88–102.
69. Ueo, H., et al., *Circulating tumour cell-derived plastin3 is a novel marker for predicting long-term prognosis in patients with breast cancer.* Br J Cancer, 2015. **112**(9): p. 1519–26.
70. Cha, S., et al., *Involvement of Cellular Prion Protein in Invasion and Metastasis of Lung Cancer by Inducing Treg Cell Development.* Biomolecules, 2021. **11**(2).
71. Cui, G., et al., *HMGB2 promotes the malignancy of human gastric cancer and indicates poor survival outcome.* Hum Pathol, 2019. **84**: p. 133–141.
72. Damaghi, M., et al., *Chronic acidosis in the tumour microenvironment selects for overexpression of LAMP2 in the plasma membrane.* Nat Commun, 2015. **6**: p. 8752.
73. Ketterer, S., et al., *Cathepsin D deficiency in mammary epithelium transiently stalls breast cancer by interference with mTORC1 signaling.* Nat Commun, 2020. **11**(1): p. 5133.
74. Zhang, X.C., et al., *Comprehensive genomic and immunological characterization of Chinese non-small

*cell lung cancer patients*. Nat Commun, 2019. **10**(1): p. 1772.
75. Morishige, M., et al., *GEP100 links epidermal growth factor receptor signalling to Arf6 activation to induce breast cancer invasion*. Nat Cell Biol, 2008. **10**(1): p. 85–92.
76. Santidrian, A.F., et al., *Nicotinamide phosphoribosyltransferase can affect metastatic activity and cell adhesive functions by regulating integrins in breast cancer*. DNA Repair (Amst), 2014. **23**: p. 79–87.
77. Glunde, K., et al., *Extracellular acidification alters lysosomal trafficking in human breast cancer cells*. Neoplasia, 2003. **5**(6): p. 533–45.
78. Zhang, Q., et al., *Mdig promotes oncogenic gene expression through antagonizing repressive histone methylation markers*. Theranostics, 2020. **10**(2): p. 602–614.
79. Nowak, R.P., et al., *First-in-Class Inhibitors of the Ribosomal Oxygenase MINA53*. Journal of Medicinal Chemistry, 2021. **64**(23): p. 17031–17050.

# 2 Tumor Microenvironment and Epigenetic Implications in Breast Cancer Progression

Mahsa Darbandi and Igor L. Bado

**Abstract**

Breast cancer (BC) poses significant challenges, driven by its diverse nature and intricate dynamics. Epigenetic modifications, such as DNA methylation, histone modifications, and noncoding RNAs, have emerged as key regulators of gene expression and BC metastasis plasticity or therapeutic resistance. Targeting epigenetic regulators and pathways associated with therapeutic resistance holds promise for overcoming treatment obstacles and enhancing treatment efficacy.

**Keywords**

Breast cancer · Epigenetics · Metastasis plasticity · Therapeutic resistance

## Abbreviations

| | |
|---|---|
| BC | breast cancer |
| CAFs | cancer-associated fibroblasts |
| MCP-1 | CCL2 |
| CTCs | circulating tumor cells |
| ctDNA | Circulating Tumor DNA |
| CSCs | cancer stem cells |
| SDF-1 | CXCL12 |
| IL-8 | CXCL8 |
| DCs | dendritic cells |
| ECM | extracellular matrix |
| EMT | epithelial-to-mesenchymal transition |
| ER | estrogen receptor |
| HER2 | human epidermal growth factor receptor 2 |
| HIFs | hypoxia-inducible factors |
| HR | hormonal receptor |
| ILCs | innate lymphoid cells |
| ILs | interleukins |
| MET | mesenchymal-epithelial transition |
| MDSCs | myeloid-derived suppressor cells |
| MMPs | matrix metalloproteinases |
| MRI | magnetic resonance imaging |
| NK | natural killer |
| NGS | next-generation sequencing |
| ncRNAs | noncoding RNAs |
| pDCs | plasmacytoid dendritic cells |
| PARP | poly-ADP ribose polymerase |
| PDGF | platelet-derived growth factor |
| P-gp | P-glycoprotein |
| PR | progesterone receptor |
| Tregs | regulatory T cells |
| TAMs | tumor-associated macrophages |
| TCR | T-cell receptor |
| TGF-beta | transforming growth factor-beta |
| TEPs | tumor educated platelets |

M. Darbandi · I. L. Bado (✉)
Department of Oncological Sciences, Tish Cancer Institute, Icahn School of Medicine at Mount Sinai, New York City, NY, USA
e-mail: igor.bado@mssm.edu



S. Bhatnagar (ed.), *Epigenetic Mechanisms in Breast Cancer Therapy and Resistance*, Advances in Experimental Medicine and Biology 1465, https://doi.org/10.1007/978-3-031-66686-5_2

| | |
|---|---|
| TILs | tumor-infiltrating lymphocytes |
| TME | tumor microenvironment |
| TNF-alpha | tumor necrosis factor-alpha |
| VEGF | vascular endothelial growth factor |

## 2.1 Introduction

### 2.1.1 Breast Cancer

Breast cancer (BC) is a heterogeneous disease with multiple subtypes based on histological (ER, PR, Her2) and molecular (luminal A/B, HER2-enriched, basal-like/TNBC, and normal breast-like subtypes) classifications (Field 1). Various factors influence the selection of standard therapeutic approaches, encompassing tumor size, morphology, grade, metastases, and expression levels of ER, PR, and HER2 (reference).

Our understanding of therapeutic resistance in BC has significantly evolved due to rapidly expanding technological advances. Surprisingly, major therapeutic targets have remained the same for decades. The most classic examples are ER and Her2 inhibitors. While early detection and treatment provide better survival outcomes, preventing local and metastatic relapses has remained challenging.

Metastasis is the process of cancer cells disseminating from ductal and lobular epithelial layers of the mammary gland to distant sites. Cancer cells are exposed to various levels of physical, immunological, and therapeutic stresses throughout this dissemination to distant organs. Only the most resistant cells successfully seed and initiate metastatic lesions at distant sites. Although BC cells can disseminate to various sites, bone, liver, lung, and brain have remained the prevalent sites of metastasis. Several mechanisms have been identified over the years, with many showing tissue-specific properties. Clinical evidence also indicates that organ-specific responses can be observed in response to targeted (e.g., anti-ER, anti-Her2) or nontargeted (e.g., chemotherapy, radiotherapy) treatment. These observations have strengthened scientific convictions about the strategic role of the tumor microenvironment. Considering the high variations between the building units of each organ, we have realized that the metastasis site matters more than expected. These sites are instrumental for metastatic progression and provide a protective milieu against major therapies. These observations align with the well-established notion of organotropism, where cancer cells preferentially colonize some organs over others. For instance, the higher frequency of bone metastasis in luminal BC over visceral metastasis further supports a strong bond between the intrinsic features of cancer cells and the site of metastasis. That said, it is still debatable whether certain concepts of metastasis, including organotropism, remain valid in advanced stages of BC metastasis.

Current therapeutic strategies for metastatic BC rely on combination therapies with the intent of restoring sensitivity to the standard of care via additive or synergistic effects. Unfortunately, most treatment options are more palliative than curative, especially in the advanced stages of the disease. Here, we explore epigenetic implications in BC metastasis, emphasizing the tumor microenvironment and therapeutic responses.

### 2.1.2 Genetic Contributions to Metastasis

Before diving into epigenetic contributions in BC metastasis, it is essential to recognize the critical role of genomic mutations. Numerous genomic investigations have identified common mutations associated with cancer initiation, progression, and therapeutic resistance. While most mutations can be sporadic, many mutations have been established as BC risk factors [1, 2]. These mutations persist in family lineages, leading to a substantially higher risk of developing BC [3–5]. This is exemplified by the well-established "Breast Cancer" genes, BRCA1 and BRCA2, that create proteins involved in fixing harmed DNA and inhibiting tumor development. Mutations in these components greatly elevate the likelihood of developing breast and ovarian cancers. Other genes, such as TP53 and PTEN, have been linked to an augmented risk of BC [3].

**Table 2.1** Common driver mutations associated with metastatic breast cancer

| | Gene | Observed mainly in | References |
|---|---|---|---|
| 1 | ABCA13 | Promoting metastasis | [6] |
| 2 | BCL6B | Exclusively in metastases | [6] |
| 3 | BMP2K | Metastasis | [7] |
| 4 | BRCA1 | Developing breast cancer and ovarian cancer | [8] |
| 5 | BRCA2 | Developing breast cancer and ovarian cancer | [8, 9] |
| 6 | CCND1 | Metastatic | [9] |
| 7 | CREBBP | Promoting metastasis | [6] |
| 8 | CTGF | Bone metastasis | [10] |
| 9 | CXCR4 | Metastasis | [10] |
| 10 | DCC | Metastasis | [6] |
| 11 | ERBB2 | Metastasis | [11] |
| 12 | ESR1 | Metastasis | [11] |
| 13 | FGFR1 | Metastasis | [9] |
| 14 | IL11 | Promoting bone metastasis | [10] |
| 15 | JAK2 | Metastasis | [11] |
| 16 | MMP-1 | Promoting bone metastasis | [10] |
| 17 | NCOR1 | Metastasis | [9] |
| 18 | NF1 | Metastasis | [11] |
| 19 | PALB2 | Metastasis | [11] |
| 20 | PIK3CA | Metastasis | [9] |
| 21 | RHC | Metastasis | [7] |
| 22 | SLC22A4 | Metastasis | [7] |
| 23 | TIAM2 | Metastasis | [6] |
| 24 | TP53 | Metastasis | [9] |
| 25 | ZNF185 | Metastasis | [6] |

In contrast, sporadic mutations can have a powerful prognostic value and often expand in response to therapeutic pressure. Table 2.1 presents a comprehensive list of common sporadic and hereditary mutations in BC. Several of these have been shown to promote genomic instability and BC progression via tumor proliferation, survival, immune escape, and metastasis mechanisms.

### 2.1.3 Change of Paradigms Pointing Toward Epigenetic Changes in Metastasis

Genetic mutations can impact epigenetic factors and alter normal homeostasis in cancer cells. As shown in Table 2.2, several epigenetic factors, including DNMTs, EZHs, HDACs, and KDMs, can be mutated in BC. Interestingly, the mutational burden of epigenetic factors (below 1% in most cases) is not comparable to well-established driver mutations such as TP53 or PIK3CA, which are above 50% [14]. Hence, the majority of epigenetic factors may impact BC reprogramming via non-genomic mechanisms. As such, relying on genetic mutations to explain the complexity of the metastatic process and its associated therapeutic resistance has been insufficient to predict and overcome metastasis in BC [12].

Emerging evidence indicating non-genomic implications in cancer reprogramming has shed light on the critical role of epigenetic mechanisms in BC development and progression and revealed epigenetics factors as promising therapeutic targets in cancer [13]. Still, there is an unmet need to better understand the crosstalk between tumor microenvironment and epigenetic states to tailor better therapeutic strategies espe-

**Table 2.2** Genes related to epigenetic regulation in pan-cancers, including BC: Specific frequencies were obtained from the AACR Project Genie [14]

| | Gene | Common genetic alterations impacting epigenetic factors | Pathways |
|---|---|---|---|
| 1 | ASXL1 | Mutation (2.62%), nonsense (0.65%), amplification (0.67%), R693* (0.10%), Y591* (0.07%) | Chromatin remodeling/DNA methylation |
| 2 | ATRX | Mutation (4.72%), frameshift (0.99%), nonsense (0.85%), loss (0.21%), fusion (0.09%) | DNA damage/repair |
| 3 | BAP1 | Mutation (1.48%), nonsense (0.32%), loss (0.28%), fusion (0.20%), amplification (0.04%) | None |
| 4 | CARM1 | Mutation (0.60%), amplification (0.42%), loss (0.10%), R419W (0.02%), A590T (0.02%) | Histone methylation |
| 5 | CIITA | Fusion (0.05%), amplification (0.07%), R2H (0.06%), loss (0.06%), E26D (0.05%) | Histone acetylation |
| 6 | DNMT1 | Mutation (1.39%), amplification (0.38%), loss (0.07%), E432K (0.02%), R1456W (0.01%) | DNA methylation |
| 7 | DNMT3A | Mutation (2.95%), R882H (0.41%), nonsense (0.38%), R882C (0.21%), loss (0.08%) | Chromatin remodeling/DNA methylation |
| 8 | DNMT3B | Mutation (1.07%), amplification (0.69%), E674K (0.02%), A241T (0.01%), loss (0.01%) | DNA methylation |
| 9 | DOT1L | Mutation (2.02%), loss (0.23%), amplification (0.06%), R853C (0.02%), T735M (0.02%) | Histone methylation |
| 10 | EP300 | Mutation (3.42%), amplification (0.12%), loss (0.10%), D1399N (0.07%), Q2268del (1.00%) | Histone acetylation |
| 11 | EZH1 | Mutation (0.69%), loss (0.06%), amplification (0.05%), R557C (0.02%), G624R (0.01%) | Histone methylation |
| 12 | EZH2 | Mutation (1.20%), c.1864-c.2195 missense (0.44%), exon 16 mutation (0.26%), codon 641 missense (0.22%), nonsense (0.13%) | Chromatin remodeling/DNA methylation |
| 13 | FBXW7 | Mutation (2.94%), R465C (0.30%), R465H (0.27%), R505C (0.23%), R479Q (0.19%) | None |
| 14 | HDAC1 | Mutation (0.24%), loss (0.09%), T322M (0.19%), C151Y (0.08%), D130N (0.08%) | None |
| 15 | HDAC2 | Mutation (0.52%), G301E (0.26%), H29N (0.26%) | Histone acetylation |
| 16 | HDAC9 | Mutation (1.83%), amplification (2.19%), R679Q (0.10%), A856D (0.18%), D957N (0.13%) | Histone acetylation |
| 17 | IDH1 | Mutation (2.75%), codon 132 missense (2.29%), R132H (1.45%), R132C (0.62%), R132G (0.12%) | Chromatin remodeling/DNA methylation, metabolic signaling |
| 18 | IDH2 | Mutation (1.14%), exon 4 mutation (0.79%), codon 140 missense (0.48%), R140Q (0.52%), codon 172 missense (0.32%) | Chromatin remodeling/DNA methylation, metabolic signaling |
| 19 | KAT6A | Amplification (1.15%), EP300 fusion (0.17%), CREBBP fusion (0.14%), fusion (0.21%), Q1667H (1.56%) | Histone acetylation |
| 20 | KDM2B | Mutation (3.00%), K930E (0.42%), D168H (0.17%), L466P (0.17%), A217T (0.08%) | Histone methylation |
| 21 | KDM3A | Mutation (0.95%), amplification (0.25%), A1103G (0.04%), E355K (0.04%), G1153E (0.04%) | Histone methylation |
| 22 | KDM6A | Mutation (2.13%), loss (0.29%), amplification (0.16%), Q555* (0.05%), fusion (0.11%) | Chromatin remodeling/DNA methylation |
| 23 | KMT2B | Mutation (3.53%), amplification (1.09%), loss (0.05%), R2587* (0.04%), A5del (0.03%) | Histone methylation |
| 24 | KMT2D | Mutation (8.24%), amplification (0.10%), loss (0.07%), fusion (0.20%), GPS2 fusion (0.20%) | Chromatin remodeling/DNA methylation |
| 25 | MGEA5 | – | Histone acetylation |
| 26 | MRE11A | Mutation (0.93%), amplification (0.11%), loss (0.12%), nonsense (0.09%), R633Q (0.02%) | Chromatin remodeling/DNA methylation |
| 27 | MSH2 | Mutation (1.55%), nonsense (0.30%), loss (0.07%), amplification (0.05%), E580* (0.04%) | Chromatin remodeling/DNA methylation |
| 28 | MSH6 | Mutation (1.86%), amplification (0.06%), fusion (0.14%), T1219I (0.05%), loss (0.05%) | Chromatin remodeling/DNA methylation |
| 29 | NCOA1 | ALK fusion (0.08%), P518R (0.77%), T377N (0.77%), A715T (0.22%), C784W (0.85%) | Histone acetylation |

| | | | |
|---|---|---|---|
| 30 | NCOA3 | Amplification (1.03%), loss (0.02%), Q1273_Q1276del (0.02%), L642* (0.01%), Q1276dup (0.02%) | Histone acetylation |
| 31 | NPM1 | Mutation (0.34%), exon 12 mutation (0.11%), amplification (0.10%), L287delinsPV (2.19%), loss (0.05%) | DNA damage/repair |
| 32 | PER1 | A912T (0.43%), E819K (3.33%), P474L (0.28%), A333V (0.20%), A943P (0.22%) | Histone acetylation |
| 33 | PRDM1 | Mutation (1.31%), amplification (0.40%), loss (0.28%), nonsense (0.05%), S367F (0.05%) | Histone methylation |
| 34 | SETD2 | Mutation (3.57%), -NF1 fusion (0.37%), loss (0.08%), X1572_splice (0.03%), fusion (0.10%) | Chromatin remodeling/DNA methylation |
| 35 | SETD7 | Amplification (0.07%), mutation (0.07%), loss (0.05%), T75S (0.05%), D88H (0.02%) | Histone methylation |
| 36 | SETDB1 | Amplification (6.01%), loss (0.07%), F670L (0.05%), N172S (0.08%), A1024T (0.04%) | Histone methylation |
| 37 | SMARCA4 | Mutation (3.78%), amplification (0.28%), loss (0.10%), T910M (0.09%), R1192H (0.04%) | None |
| 38 | SMARCB1 | Mutation (0.68%), loss (0.21%), amplification (0.19%), R377H (0.09%), K364del (0.04%) | None |
| 39 | SMYD3 | Amplification (0.43%), loss (0.04%), S154L (0.02%), A252V (0.01%), R165* (0.01%) | Histone methylation |
| 40 | STAT3 | Mutation (1.03%), exon 21 mutation (0.17%), exon 20 mutation (0.13%), amplification (0.10%), loss (0.05%) | JAK/STAT signaling |
| 41 | TET2 | Mutation (3.76%), c.3310-c.4434 missense (1.00%), nonsense (0.97%), c.3400-c.4331 missense (0.92%), c.5533-c.6005 missense (0.40%) | Chromatin remodeling/DNA methylation |
| 42 | WHSC1L1 | Amplification (2.52%), loss (0.35%), T1271M (0.02%), fusion (0.06%), NUTM1 fusion (0.06%) | Histone methylation |

cially in the context of BC metastasis [1, 2]. Additionally, the translation of epigenetic investigations remains limited due to high risks of toxicity as a result of the ubiquitous expression of these factors. With a more thorough assessment of the appropriate dosing, timing of drug delivery, and the right combination, we may establish better therapeutic regimens for long-term responses with minimal toxicity for BC patients.

## 2.2 Principles of Epigenetic Regulation

Epigenetic regulation plays a pivotal role in cellular differentiation and serves as an additional layer of gene control, surpassing the DNA sequence in terms of malleability, and underscoring its significance in cancer development and cellular heterogeneity [15]. Epigenetic modifications, including DNA methylation, histone modifications, noncoding RNAs (ncRNAs), chromatin remodeling, and genomic imprinting, can alter the accessibility of genes, either by promoting or inhibiting their expression (Fig. 2.1).

### 2.2.1 DNA Methylation

DNA methylation involves adding a methyl group to DNA molecules, typically at cytosine residues in CpG dinucleotides, through enzymatic activities of DNA methyltransferases (DNMTs). Major DNMTs are DNMT1, DNMT3A, DNMT3B, and DNMT3L. An increased expression of DNMT1 is observed in many types of cancers, and this overexpression is associated with cellular transformation, while reduced DNMT1 expression levels seem to be associated with a protective function. DNMT3B is overexpressed in BC tissues compared to DNMT1 and DNMT3A [16] and was shown to promote BC metastasis [17]. Although DNA methylation is a dynamic process, aberrant DNA methylation patterns can lead to the silencing of tumor-suppressor genes or activating oncogenes, promoting uncontrolled cell growth and division [16]. DNA demethylation is facilitated by a group of enzymes known as ten-eleven translocation methylcytosine dioxygenases or TETs. These enzymes, namely, TET1, TET2, and TET3, convert 5mC into 5-hydroxymethylcytosine (5-hmC) through hydroxymethylation. Their role in DNA demethylation allows for the reactivation of genes previously silenced by DNMTs [18]. BC involves additional proteins with DNA demethylase capabilities, including the growth arrest and DNA-damage inducible protein (GADD45), as well as the cytidine deaminases family of proteins, which include activation-induced cytidine deaminase (AID) and apolipoprotein B mRNA editing catalytic polypeptide-like family (APOBEC) [15]. GADD45A, in particular, exhibits strong connections between DNA repair and the regulation of genes through epigenetic mechanisms. In BC, GADD45 is linked to BRCA1, potentially impacting disease development via nucleotide excision repair [15]. Abnormal GADD45A methylation is observed in BC [19]. AID proteins help active DNA demethylation, converting 5-mC to thymine through deamination. AID is essential for epithelial-to-mesenchymal transition (EMT) in mammary epithelial cells. APOBEC1 exhibits DNA demethylase activity, affecting tumor evolution in ER+/HER2- BC [20]. APOBEC-induced mutagenesis inhibits breast tumor growth, eliciting immune responses [21].

### 2.2.2 Chromatin Modifications

Histones are proteins that compactly package DNA into chromatin and can undergo posttranslational modifications such as acetylation, methylation, phosphorylation, and ubiquitination. These modifications, often called the "histone code," serve as a language that regulates gene expression by influencing chromatin structure and the recruitment of transcription factors [22, 23]. Histone acetylation involves adding acetyl groups to histones H3 and H4, a process facilitated by histone acetyltransferases (HATs). Conversely, these acetyl groups are eliminated by enzymes known as histone deacetylases (HDACs). Notable HAT enzymes encompass

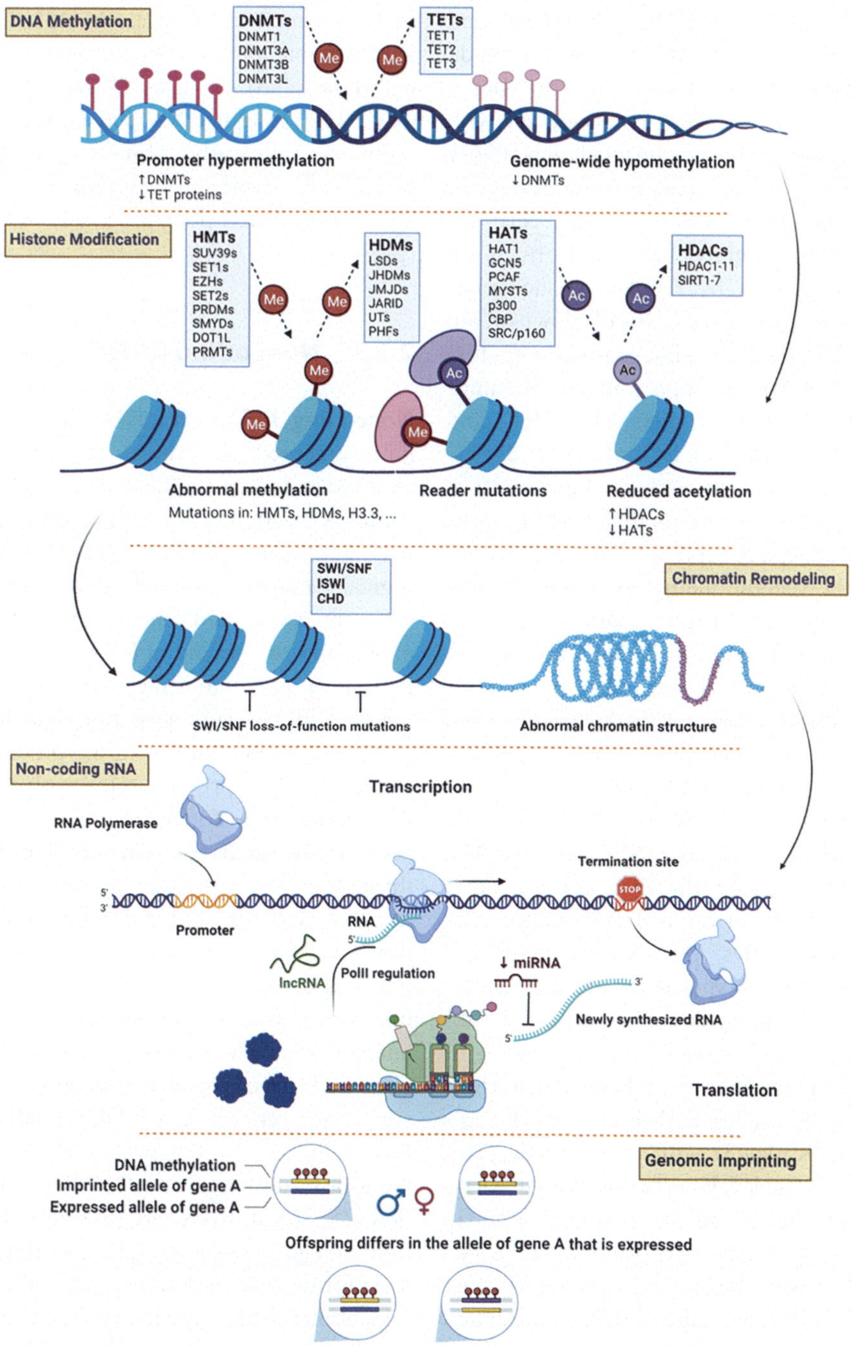


**Fig. 2.1** Epigenetic factors in BC. The control of gene expression in cancer through epigenetic regulation is a complex process involving various modifications such as DNA methylation, histone modifications, noncoding RNAs (ncRNAs), chromatin remodeling, and genomic imprinting. These modifications play a crucial role in influencing cancer development and the diversity of cells within tumors. Key players in each epigenetic process and their implications in BC are depicted

HAT1, GCN5, PCAF, MYSTs, p300, CBP, and SRC/p160, whereas HDAC enzymes include HDAC1 to HDAC11, as well as SIRT1 to SIRT7. In BC, P300 overexpression is associated with higher cancer risk and lower survival rates. P300/CBP enzymes also regulate proliferation, cell death, EMT, and metastasis in BC [23]. Sirtuins, a class III HDAC family, have dual effects on breast carcinogenesis by regulating oncogenes and tumor-suppressor genes. SIRT1, downstream of BRCA1, negatively regulates Survivin via histone H3 deacetylation, impacting breast tumor growth. Other Sirtuin members, like SIRT2 and SIRT7, are implicated in BC, affecting TNBC aggressiveness and elevated in human BCs, respectively [23]. Histone methylation at lysine residue can activate or repress transcription, contingent on the specific methylation site, thereby establishing diverse histone marks/code [24]. Histone lysine and arginine residues can undergo mono-, di-, or trimethylation. Histone methyltransferases (HMTs) transfer methyl groups from SAM to lysine's ϵ-amino group, while demethylases (HDMs) regulate methylation reversibility [22]. Enzymes classified as HMTs include SUV39s, SET1s, EZHs, SET2s, PRDMs, SMYDs, DOT1L, and PRMTs. On the other hand, enzymes categorized as HDMs encompass LSDs, JHDMs, JMJDs, JARID, UTs, and PHFs. In BC, luminal A subtypes show elevated acetylation and methylation compared to basal subtypes. Investigators explored the correlation of seven histone protein markers (H3K18ac, H3K9ac, H4R3me2, H3K4me2, H4K12ac, H4K16ac, H4K20me3) with tumor grade [22]. Reduced H4K16ac levels were observed in early-stage cancer, indicating its potential role in pathogenesis. Decreased H4R3me2, H3K9ac, and H4K16ac were linked to larger tumor size, while high H4R3me2 and H3K9ac correlated with a lower lymph node stage. Different BC subtypes exhibited distinct H3K4 acetylation and methylation dynamics. PI3K/AKT signaling regulated H3K4 methylation, with elevated H3K4me3 levels associated with breast tumors. H3K27ac played a significant role in BC progression and EMT. Loss of repressive mark H3K27me3 indicated poor prognosis. Enrichment of H3K27me3 in FOXC1, RAD51, CDH1, and RUNX3 gene promoters contributed to enhanced cell growth and metastasis [22]. Hypermethylation of CDH1 through DNA and H3K27 trimethylation is frequently observed in metastasis. H4K20me3 exhibited a significant decrease in BC and predicted poor prognosis independently [22].

### 2.2.3 Noncoding RNAs

**Noncoding RNAs** (ncRNAs), like microRNAs, have emerged as critical regulators of gene expression and are implicated in BC. MicroRNAs (miRNAs) and long noncoding RNAs (lncRNAs) play significant roles in modulating various biological processes, including cell proliferation, differentiation, and metastasis. Dysregulated expression of specific ncRNAs has been identified as a potential diagnostic and prognostic marker for BC subtypes. miRNAs function as compact guiding entities in the process of RNA silencing. Their primary role involves the negative modulation of gene expression, operating at both the mRNA and protein tiers. This is achieved through mechanisms such as mRNA target degradation and translational inhibition. A single miRNA often targets several genes, and its regulatory effect is tissue specific [25, 26]. In cancer, miRNAs play a role in oncogenesis, metastasis, and resistance to various therapies and can be classified as oncogenes (oncomirs) or tumor-suppressor genes [25, 26]. Additionally, both pro-metastatic ("metastamiRs") and metastasis-suppressor miRNAs have been identified [25, 26]. miRNAs are present in the circulation of cancer patients and can potentially be used for disease monitoring, as well as new biological therapeutic agents, targets, or biomarkers for patient-tailored BC treatment. In invasive BC, the lncRNA H19 exhibits abnormal hypermethylation, suggesting tumor-suppressive functions. HOTAIR, another lncRNA, can recruit writer proteins (MLL1, MLL3, P300/CBP) to its promoter, increasing histone acetylation and H3K4me3 levels. This suppresses apoptosis and drives BC progression [27].

### 2.2.4 Chromatin Remodeling

Chromatin is organized in a three-dimensional structure within the nucleus, which plays a crucial role in gene regulation [28]. Alterations in chromatin organization, such as changes in topological domains, chromatin looping, and enhancer-promoter interactions, have been observed in human cancers [29]. The fundamental units of chromatin, nucleosomes, have a dynamic role where their repositioning impacts DNA accessibility, leading to gene expression modulation. Chromatin remodeling complexes utilize the energy derived from ATP hydrolysis to alter nucleosome position, composition, or structure, thereby regulating gene accessibility. These complexes can be classified into different families, such as SWI/SNF (SWItch/Sucrose Non-Fermentable), ISWI (Imitation SWI/SNF), and CHD (Chromodomain helicase DNA-binding), based on their subunit composition and enzymatic activity [30]. In BC, altered chromatin remodeling can lead to abnormal gene regulation, promoting tumor initiation and progression. Dysregulation of these complexes contributes to the development of drug resistance and has implications for targeted therapies.

### 2.2.5 Genomic Imprinting

Genomic imprinting is an epigenetic phenomenon involving selective silencing or activation of genes based on their parental origin. Alterations in imprinting patterns have been associated with BC development. Abnormalities in imprinted genes can disrupt critical cellular pathways, contributing to tumorigenesis and treatment resistance. Studies have identified several putatively imprinted genes in healthy breast tissue, including HM13, IGF-2, and PHLDA2 [31, 32]. In BC, HM13 is featured by significant loss of imprinting and expression upregulation, which is linked to DNA demethylation [32]. Other imprinted genes typically demonstrate lower expression in cancer, often associated with copy number variation and aberrant DNA methylation [32].

## 2.3 Epigenetic Mechanisms in BC

The impact of epigenetic factors on tumor outcomes remains complex and context specific. The downstream effectors, which often involve transcription factors, determine tumor progression outcome, depending on their oncogenic or tumor-suppressive functions [15, 24, 33]. Many cancer-associated genes are modulated epigenetically to facilitate BC progression and metastasis (Table 2.3). In metastasis, DNA methylation and histone modifications can regulate the expression of EMT-inducing transcription factors, such as Snail, Twist, and ZEB1 [34]. Together with other EMT markers, such as E-cadherin, N-cadherin, and vimentin, these factors are crucial for the initiation and progression of metastasis [34]. BC metastasis suppressor 1 (BRMS1), shown to regulate focal adhesions, growth factor receptor, and NF-KB signaling processes, is often silenced through DNA hypermethylation and histone modifications in BC cells [35]. Similar epigenetic alterations impact several miRNAs involved in metastasis. As metastasis suppressors, miR-200 family members are frequently downregulated in the metastasis setting [36]. On the other hand, lncRNA, such as HOTAIR, has been implicated in promoting BC metastasis by recruiting chromatin-modifying complexes to specific genomic regions, leading to altered gene expression profiles [37].

DNA methylation patterns differ among BC subtypes. Interestingly, CpG analysis can capture the molecular variabilities between BC subtypes, comparably to the PAM50 analysis. Overall, this study further supports the predictive potential of epigenetic alterations in BC [38]. Differential methylation patterns in genes such as ER, PR, or HER2 can impact receptor expression and responsiveness to endocrine/hormonal therapies in breast tumors [39, 40]. This may explain why epigenetic modulators such as DNA methylation inhibitors (e.g., decitabine) and histone deacetylase inhibitors (e.g., vorinostat) have shown promise in preclinical and clinical studies for inhibiting metastasis in BC [41].

**Table 2.3** Common epigenetically regulated genes in BC

| | | | |
|---|---|---|---|
| CDH1 | Hypermethylation | Tissue | [42] |
| TMS1, BRCA1, ESR1, and PRB | Hypermethylation | Serum | [43] |
| HYAL2 | Hypomethylation | Peripheral blood | [44] |
| KIF1A promoter | Hypermethylation | Plasma | [45] |
| DOK7 promoter | Hypermethylation | Whole blood | [46] |
| ITIH5, DKK3, and RASSF1A promotor | Hypermethylation | Serum | [47] |
| ATM | Hypermethylation | White blood cells | [48] |
| Global genome | Hypomethylation | White blood cells | [49] |
| PRAC2, TDR10, and TMEM132C | Hypermethylation | Tumor tissue | [50] |
| RARβ2, RASSF1A, MINT17, and MINT13 | Hypermethylation | Tumor tissue | [51] |
| MGMT, BRCA1, MLH1, RAD9, LDLRAP1 (ARH1), GPC3, HOXD11, LAMA3, LAMB3, LAMC2, ROBO1, SYK, CCND2, 14–3-sigma (SFN), AK5, CDKN1C, CDKN2A, ER, FOXA2, PGR, RAR-β, RASSF1A, RUNX3, SFRP1, WIF1, WRN, WT1, APC, HIC1, HOXA5, DAPK, BCL2, DCC, TWIST, TMS1, and CST6 | Hypermethylation | – | [52] |
| BCSG1, CDH1, CDH3, CDH13, CAV1, NAT1, and UPA | Hypomethylation | | [52] |
| DLEC1, GRIN2B, HOXA1, MT1G, SFRP4, TMEFF2, APC, GSTP1, HOXA10, IGF2, RARβ, RASSF1A, RUNX3, HIN-1, and SFRP1 | Hypermethylation | Tumor tissue | [53] |
| RIL, HIN-1, RASSF1A, CDH13, and RARβ2 | Hypermethylation | Bovine tumor tissue | [54] |
| ESR1 | Hypermethylation | Tumor tissue | [55–57] |
| ER | Hypermethylation | Tumor tissue | [58] |
| PSAT1 promoter | Hypermethylation | Tumor tissue | [59] |
| BRCA1 promoters | Hypermethylation | Tumor tissue | [60, 61] |
| PTEN promoter | Hypermethylation | Tumor tissue | [62, 63] |
| HIN-1, twist, cyclin D2, and RASSF1A | Hypermethylation | Tumor tissue | [64] |
| P16INK4a, BRCA1, BRCA2, ERα and RARβ2, APC, and RASSF1A | Hypermethylation | Tumor tissue | [65] |
| RASSF1A and CCND2 | Hypermethylation in ER+ | Tumor tissue | [66] |
| PGR, TFF1, CDH13, TIMP3, HSD17B4, ESR1, and BCL2 | Hypermethylation in ER | Tumor tissue | [57] |
| ESR1, TGFBR2, PTGS2, and CDH13 | Hypermethylation in PR | Tumor tissue | [57] |
| FAM124B, ST6GALNAC1, NAV1, and PER1 | Methylation status in ER+/PR+andER−/PR− | Tumor tissue | [67] |

(continued)

**Table 2.3** (continued)

| | | | |
|---|---|---|---|
| CDH1 | Hypermethylation | Tissue | [42] |
| CCND2, RASSF1A, APC, and HIN1 | Hypermethylation | Tumor tissue | [68] |
| ITIH5, DKK3, and RASSF1A | Hypermethylation | Serum samples | [47] |
| APC, BIN1, BMP6, BRCA1, CST6, ESR2, GSTP1, P16, P21, and TIMP3 | Hypermethylation | Tumor tissue | [69] |
| CST6 | Hypermethylation | Plasma | [70] |
| BMI2, ZEB1, ZEB2, and sec-23a | miR-200 | Tumor tissue | [71, 72] |
| TP53 or ESR1 | miR-145 | Tumor tissue | [73] |
| AMOT, ERBB3, VEGF-A, HMGB, YAP1, E2F1, PTEN, LAMC1, ZEB1/ZEB2/SIP1, ITGA5, Notch2, FGF2, ERBB2, and Bcl-w | miR-205 | Tumor tissue | [74] |
| MMP11 and ARID3B | miR-125b | Tumor tissue | [75, 76] |
| NDRG2 | miR-181a | Tumor tissue | [77] |
| SMAD7 | miR-106b-25 | Tumor tissue | [78] |
| FZD3, ITGA5, M-RIP, MMP16, RDX, and RHOA | miR-31 | Tumor tissue | [79] |
| H-RAS, HMGA2, LIN28, and PEBP1 | Let-7 | Tumor tissue | [80–82] |
| ER | miR-375 | Tumor tissue | [83] |
| HOXD10 and RhoC | miR-10b | Tumor tissue | [84–86] |
| BCL-2, PTEN, PDCD4, TPM1, and SERPINB5 | miR-21 | Tumor tissue | [87–89] |
| SOX4 and tenascin-C | miR-335 | Tumor tissue | [90–92] |
| RHOC | miR-10b | Tumor tissue | [86] |
| ESR1 | miR-206 | Tumor tissue | [93] |
| ERBB2 and ERBB3 | miR-125 | Tumor tissue | [94–96] |
| TP53, CCND1, CDK4, and CDK6 | miR-34 | Tumor tissue | [97–99] |
| TP53 | miR-155 | Tumor tissue | [100, 101] |
| BAK1 | miR-125b | Tumor tissue | [102] |
| ERBB3, VEGF, and HIF-1 | miR-148 | Tumor tissue | [103] |
| VEGF | miR-152 | Tumor tissue | [103] |

## 2.4 Epigenetic Drugs in BC

To target epigenetic alterations in cancer, multiple inhibitors have been developed. In the context of BC, there are two major classes of epigenetic drugs commonly used: DNA methyltransferase inhibitors (DNMT inhibitors) and histone deacetylase inhibitors (HDAC inhibitors), while other drugs (e.g., EZH2 inhibitors) application to BC has not been implemented yet. The frequently utilized categories of epigenetic drugs in BC have been elaborated upon in Table 2.4.

## 2.5 Crosstalk Between Cancer Cells and TME

The tumor microenvironment (TME) encompasses a diverse array of cellular and noncellular components that surround the tumor, including stromal cells, immune cells, fibroblasts, endothelial cells, and the extracellular matrix (ECM) [104–106]. Within the context of BC, the TME plays a pivotal role in promoting tumor growth, invasion, angiogenesis, immune evasion, and metastasis [104]. The intricate interactions between tumor cells and the TME lead to bidirectional communication that profoundly influences the behavior of both the tumor cells and the surrounding cells [106, 107].

### 2.5.1 From Cancer Cells to TME

Tumor cells have the ability to release various factors, including cytokines, growth factors, and chemokines, which exert significant effects on tumor growth, invasion, angiogenesis, immune evasion, and metastasis [108, 109]. These factors can also recruit and activate different immune cells, such as macrophages, T cells, and regulatory cells, which can have both pro- and antitumor effects depending on their activation status and the signals they receive from the TME [108, 109]. Inflammatory signals, such as interleukins (ILs) and tumor necrosis factor-alpha (TNF-α), can activate signaling pathways that influence DNA methylation patterns and histone modifications. These changes can result in altered gene expression profiles that promote tumor progression, immune evasion, and therapy resistance. Inflammatory signaling can also impact the expression of ncRNAs, including miRNAs, which regulate the expression of target genes involved in tumor cell behavior [108]. Among the factors secreted by tumor cells, TNF-α plays a role in promoting inflammation, inducing programmed cell death (apoptosis), and stimulating angiogenesis through the production of vascular endothelial growth actor (VEGF) and other angiogenic factors [107]. Transforming growth factor-beta (TGF-beta) exhibits both tumor-promoting and

**Table 2.4** Epigenetic drugs with potential clinical translation in BC

| Drug | Drug categories | Drug target |
|---|---|---|
| Vorinostat | Histone deacetylase/HDAC inhibitors | HDAC1, HDAC2, HDAC3, HDAC6 |
| Entinostat | Histone deacetylase/HDAC inhibitors | HDAC1, HDAC2, HDAC3 |
| Romidepsin | Histone deacetylase/HDAC inhibitors | HDAC1, HDAC2, HDAC3, HDAC8 |
| Azacitidine | Histone deacetylase/HDAC inhibitors | HDAC1, HDAC2, HDAC3 |
| Molibresib | BET inhibitor | BRD2, BRD3, BRD4, BRDT |
| Decitabine | Histone hypomethylation | DNMTs |
| Belinostat | Histone deacetylase/HDAC inhibitors | HDAC1, HDAC10, HDAC11, HDAC2, HDAC3, HDAC4, HDAC5, HDAC6, HDAC7, HDAC8, HDAC9 |
| Panobinostat | Histone deacetylase/HDAC inhibitors | HDAC1, HDAC10, HDAC11, HDAC2, HDAC3, HDAC4, HDAC5, HDAC6, HDAC7, HDAC8, HDAC9 |
| Chidamide | Histone deacetylase/HDAC inhibitors | HDAC1, HDAC10, HDAC2, HDAC3 |

tumor-suppressing effects, acting as a tumor-suppressor in the early stages of cancer but facilitating invasion, metastasis, and immune evasion in advanced stages [110]. VEGF promotes angiogenesis, ensuring a sufficient blood supply to the growing tumor and supporting its progression and metastasis [111]. ILs such as IL-6, IL-8, and IL-10 have diverse effects on tumor progression, with IL-6 promoting tumor growth and metastasis, IL-8 involved in angiogenesis and immune cell recruitment, and IL-10 exhibiting immunosuppressive properties [112]. Chemokines like CCL2 (MCP-1), CXCL12 (SDF-1), and CXCL8 (IL-8) play crucial roles in immune cell recruitment and migration, potentially promoting inflammation, tumor growth, and metastasis [113–115]. Platelet-derived growth factor (PDGF) contributes to tumor growth, angiogenesis, and stromal cell recruitment by stimulating the proliferation of vascular smooth muscle cells, fibroblasts, and pericytes [116]. Breast tumor cells downregulate antigen expression that would typically make them recognizable to immune cells. Additionally, they release immunosuppressive factors and recruit immune regulatory cells such as regulatory T cells or myeloid-derived suppressor cells, suppressing immune responses [117, 118]. Metastatic cancer cells can evade immune surveillance, enabling their survival and growth at distant sites. Immune evasion is mediated by the modulation of genes involved in immune response regulation, including immunosuppressive factors (e.g., PD-L1, TGF-β, and IL-10), and the downregulation of major histocompatibility complex (MHC) molecules. These molecular changes hinder the recognition and elimination of cancer cells by the immune system. It's important to note that the secretion of these factors can vary among different types of tumors and even within different stages of the same tumor. Figure 2.2 provides a summary of the immune evasion process in cancer.

### 2.5.2 From TME to Cancer Cells

The TME in BC relies on the presence and functions of stromal cells, which include endothelial cells, fibroblasts, and the ECM [106, 119]. Both tumor and stromal cells, including fibroblasts, actively deposit and modify the ECM, creating an environment that promotes tumor growth and invasion [106, 120]. These stromal elements collectively contribute to tumor growth, angiogenesis, invasion, and the modulation of immune responses within the BC microenvironment [106, 120]. Endothelial cells, which line the blood vessels, have a crucial function in angiogenesis or forming new blood vessels [106, 120]. In BC, angiogenesis is vital for tumor growth and metastasis [106, 120]. Tumor cells release factors, including VEGF, to stimulate the development of new blood vessels that supply nutrients and oxygen to the tumor. Disrupting angiogenesis and targeting the formation of new blood vessels are strategies employed in anticancer therapies [106, 120]. In BC, fibroblasts can undergo phenotypic changes and transform into cancer-associated fibroblasts (CAFs) [106, 119]. These CAFs actively contribute to tumor progression by secreting various factors, such as growth factors, cytokines, and chemokines [106, 119]. CAFs can also influence tumor growth, invasion, immune response, and angiogenesis through the secretion of EMC, growth factors and matrix remodeling enzymes [121, 122]. CAFs can secrete factors that promote DNA methylation changes and histone modifications in cancer cells, leading to alterations in gene expression and tumor cell behavior [123]. Fibroblasts play a vital role in producing and maintaining the ECM [106, 119]. The ECM is a complex meshwork of proteins, glycoproteins, and proteoglycans that provide structural support to tissues [106, 120]. The ECM can act as a physical barrier, influence cell signaling, and regulate the behavior of tumor cells [106, 120]. The composition and stiffness of the ECM can significantly impact tumor progression and response to therapy [106, 120]. ECM components and remodeling enzymes can influence epigenetic modifications, tissue architecture, and signaling in BC cells. ECM components can degrade using enzymes like matrix metalloproteinases (MMPs) and their inhibitors, tissue inhibitors of metalloproteinases (TIMPs), to invade surrounding tissues [126, 127]. The bal-

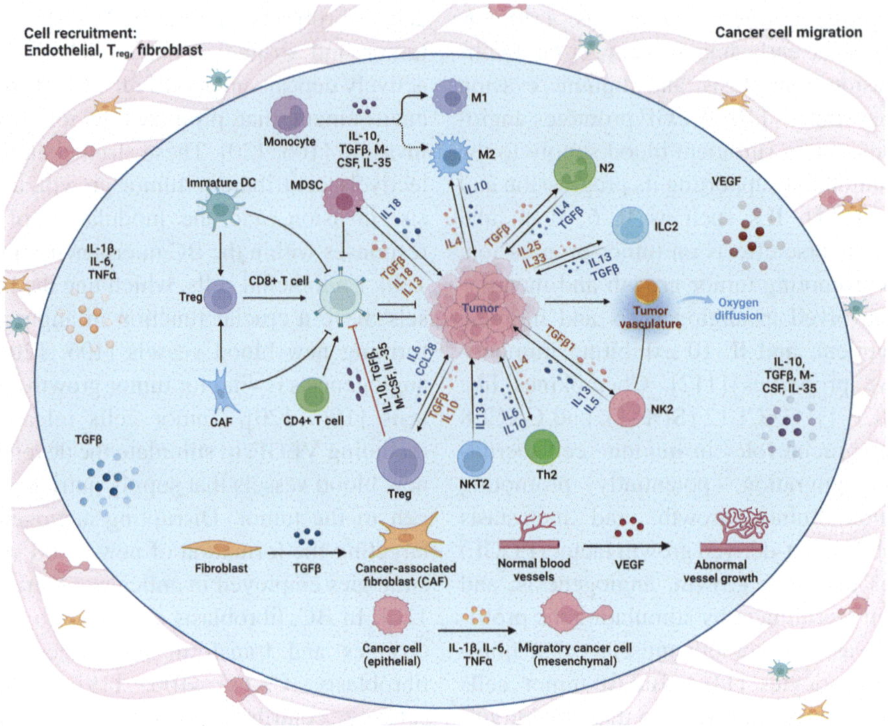


**Fig. 2.2** Signaling between tumor cells and their microenvironment. The tumor microenvironment (TME) is made up of different types of cells and other components. These include stromal cells, immune cells, fibroblasts, endothelial cells, and the extracellular matrix (ECM). The diagram provided shows how tumor cells communicate with the TME in both directions, impacting tumor growth, invasion, angiogenesis, immune evasion, and metastasis. Tumor cells release substances like cytokines, growth factors, and chemokines that affect the behavior of tumor and stromal cells. Stromal cells, such as fibroblasts, endothelial cells, and ECM components, play an active role in promoting tumor progression by releasing substances and altering the ECM. This ultimately influences gene expression, tumor cell behavior, and response to treatment. The illustration also highlights key signaling pathways and interactions within the TME

ance between MMPs and TIMPs is tightly regulated, with dysregulation favoring ECM degradation and cancer cell invasion.

Moreover, tumor cells can respond to ECM signals, altering their behavior and promoting tumor progression [124, 125]. Genes involved in cell adhesion, such as integrins and selectins, are crucial for cancer cell interactions with the ECM and endothelial cells during extravasation at distant sites. Integrin-mediated signaling pathways activated by ECM components can modulate DNA methylation, histone modifications, and miRNA expression, impacting tumor cell adhesion, invasion, and metastasis [126, 127]. Signaling pathways that regulate cytoskeletal dynamics, such as Rho GTPases and focal adhesion kinase (FAK), also play a role in cell migration and invasion [126, 127]. Targeting specific components of the TME is an area of active research and holds potential as a strategy to improve BC treatment outcomes [106, 120].

### 2.5.3 Metastasis Plasticity in BC

Breast tumor cells can enter a dormant state, remaining quiescent for extended periods without actively proliferating. Dormant tumor cells can evade therapy and later reawaken, leading to disease recurrence or metastasis. The mechanisms regulating tumor cell dormancy are still being investigated [126]. Metastasis plasticity in BC refers to the dynamic and reversible changes in the molecular and cellular characteristics of cancer cells in response to changes in the microenvironment, such as nutrient availability, oxygen levels, growth factors, and interactions with stromal and immune cells during the metastatic process. It involves the acquisition of diverse phenotypes that enable cancer cells to survive, invade surrounding tissues, enter the bloodstream or lymphatic system, evade immune surveillance, establish secondary tumors in distant organs, and acquire resistance to therapies [127]. The plasticity of cancer cells poses challenges in predicting treatment response, monitoring disease progression, and developing personalized treatment strategies [128]. Various molecular mechanisms, including genetic and epigenetic alterations, dynamic changes in gene expression, activation of signaling pathways, and interactions with the microenvironment, drive metastasis plasticity. The metastatic process encompasses several key stages, including invasion, intravasation, circulation, extravasation, and metastatic colonization. BC cells at the primary tumor site acquire invasive properties that allow them to break through the basement membrane and invade the surrounding stroma. Changes in cell adhesion molecules, secretion of matrix-degrading enzymes, and activation of signaling pathways involved in cell motility facilitate this process [127]. Cancer cells invade nearby tissues and intravasate into the bloodstream or lymphatic vessels [129]. However, most cells do not survive due to shear stress, anoikis (apoptosis triggered by the loss of cell-ECM interactions), or attacks from immune cells.

BC cells that successfully extravasate must adapt to the new microenvironment before colonizing. Phenotypic changes, including mesenchymal-epithelial transition (MET), may occur, allowing cancer cells to regain epithelial characteristics. Interactions with local stromal cells, remodeling of the extracellular matrix, and promoting angiogenesis establish a supportive niche for tumor growth [129]. It is now well-accepted that cancer cells with hybrid EMT phenotypes have more aptitude to initiate distant metastasis successfully. This is driven by the increased plasticity of cancer cells, which facilitates phenotypic switching between different EMT states. Breast tumor cells can exhibit stem cell-like characteristics, known as cancer stem cells (CSCs) or tumor-initiating cells, with self-renewal capacity and the ability to differentiate into various cell types within the tumor. CSCs are implicated in tumor initiation, progression, treatment resistance, and recurrence [130, 131]. Overall, CSC properties are essential for the metastasis process, where tumor-initiating capabilities are required.

### 2.5.4 Epigenetic Alterations in CTCs

CTCs are essential for the metastatic process. CTC clusters formed by the aggregation of multiple CTCs exhibit distinct properties compared to single CTCs. Studies have shown that CTC clustering induces epigenetic plasticity, leading to the acquisition of a distinct DNA methylation profile. CTC clusters exhibit unique DNA methylation patterns compared to single CTCs or primary tumor cells, and this plasticity enables the activation of genes involved in survival, extravasation, and colonization at metastatic sites [129, 132]. Conversely, the expression of genes associated with cell adhesion and immune surveillance may be downregulated through DNA hypermethylation. Targeting the specific DNA methylation alterations in CTC clusters may provide new opportunities for preventing or inhibiting metastasis [132]. Additionally, profiling DNA methylation patterns in CTC clusters could serve as a noninvasive biomarker for early detection of metastatic potential and monitoring treatment response [129].

## 2.6 Mechanisms of Therapeutic Resistance in BC

Therapeutic resistance refers to the ability of BC cells to evade the effects of anticancer treatments, leading to a reduced or absent response to therapy. This resistance can occur during various treatment modalities, including chemotherapy, radiation therapy, targeted therapies, and hormonal therapies [133, 134].

### 2.6.1 Epigenetic Contribution

DNA methylation-mediated silencing of tumor-suppressor genes is commonly associated with acquired resistance to chemotherapy and hormonal therapy in BC. Hypermethylation of genes involved in DNA repair, cell cycle regulation, and drug metabolism can reduce treatment sensitivity and promote resistance [135]. Methylation-mediated silencing of ER alpha (ERα) can also confer resistance to hormonal therapies targeting ERα signaling [135]. Altered histone modifications, such as increased histone deacetylation and histone methylation, have been linked to resistance in BC[135]. Alterations in HER2 expression can lead to resistance against HER2-targeted therapies like trastuzumab [136]. In HR-positive BC, resistance mechanisms may involve alterations in HR expression or signaling, crosstalk between signaling pathways, or activation of alternative survival pathways. EMT can confer resistance to chemotherapy and targeted therapies. Cells undergoing EMT may exhibit altered signaling pathways and reduced drug sensitivity, contributing to therapeutic resistance [137].

Dysregulation of ncRNAs, particularly miRNAs, has been implicated in BC resistance [135]. Altered expression of specific miRNAs can affect the expression of genes involved in drug metabolism, apoptosis, and cell proliferation, thereby influencing the response to treatment [25]. Epigenetic modifications can regulate the expression of drug transporters and efflux pumps, influencing the pharmacokinetics of anticancer drugs. Increased expression of ATP-binding cassette (ABC) transporters, such as P-glycoprotein (P-gp), can lead to decreased intracellular drug accumulation, limiting the efficacy of chemotherapy drugs [135]. BC cells can develop resistance to chemotherapy drugs by overexpressing efflux pumps, such as P-gp, which actively transport drugs out of the cells. This efflux mechanism reduces intracellular drug concentration and limits the effectiveness of chemotherapy. Targeting the epigenetic modifications and CSCs has become an area of intense research to overcome therapeutic resistance and improve treatment outcomes [138].

### 2.6.2 Tumor Heterogeneity

The intratumoral heterogeneity of BC can contribute to therapeutic resistance as not all cells within the tumor respond uniformly to treatment [133, 134, 139]. BC cells can activate various signaling pathways that promote cell survival and resistance to treatment. For example, the activation of the PI3K/AKT/mTOR pathway, which is frequently dysregulated in BC, can confer resistance to both chemotherapy and targeted therapies. Other pathways, such as the MAPK and JAK/STAT, have also been implicated in therapeutic resistance [134]. Interactions between cancer cells and the microenvironment can create a supportive niche that promotes survival and therapy resistance [106, 109].

The acidic microenvironment within tumors can influence therapeutic response [136]. Acidic conditions can affect drug uptake, alter drug-target interactions, and promote the activation of survival pathways, contributing to therapeutic resistance [136, 140]. The extracellular pH (pHe) of tumor tissues is often acidic, and acidic metabolites, such as lactic acid, caused by anaerobic glycolysis in hypoxia, seem to be the leading cause. Accumulating evidence shows that an acidic microenvironment regulates cellular phenotype [140]. Acidic pH increases not only the activation of some lysosomal enzymes with acidic optimal pH but also the expression of some genes involved with pro-metastatic factors [140]. The pH in the TME can influence the cytotoxicity of anticancer drugs. Molecules diffuse passively across the cell membrane most efficiently when the pH is near neutral. At low pH, the protonation of weakly basic drugs reduces their diffusion across the cell membrane, leading to reduced drug uptake and decreased efficacy [136]. Hypoxia can activate survival pathways, such as HIF-1α signaling, which promotes cell survival and resistance to chemotherapy and radiation therapy. Hypoxic conditions also affect the TME and impair drug delivery to cancer cells [136]. The Warburg effect is a metabolic alteration observed in cancer cells where they rely more on glycolysis (a less efficient form of energy production) even in the presence of oxygen [136].

This metabolic shift can contribute to therapeutic resistance by providing cancer cells with the necessary energy and building blocks for survival and proliferation. Dysregulation of apoptotic pathways can contribute to therapeutic resistance by preventing cancer cells from undergoing cell death in response to treatment-induced stress [136]. Altered drug metabolism and elimination can impact the effectiveness of chemotherapy drugs. Changes in drug-metabolizing enzymes and drug transporters can lead to decreased drug concentrations in cancer cells, reducing the cytotoxic effects of treatment [136]. Considering the frequent detection of dormant cells during tissue colonization, investigating metabolic changes involved in this process may provide alternative therapeutic opportunities [126].

### 2.6.3 Immune Escape

Immune checkpoint proteins, such as PD-L1, can be upregulated in BC cells, inhibiting immune responses and allowing tumor cells to escape immune surveillance [141, 142]. PD-L1 binding to PD-1 on immune cells induces inhibitory responses, which can promote immune evasion and tumor progression [141]. Most research associated with PD-L1 and PD-1 has been focused on their extrinsic role in inhibiting the immune system, but more recently, a tumor-intrinsic role of PD-L1 and PD-1 is emerging in some cancer types; however, these roles remain to be fully characterized in all cancers [141]. PD-1/PD-L1 checkpoint blockade is at the cutting edge of research, offering cancer patients hope for new treatment regimens with the potential to have substantial clinical benefits and prolong survival (Field 141, 142). The presence of minor subclones within the tumor that harbors resistance-conferring genetic alterations can lead to treatment failure and disease recurrence. Clonal evolution and selection of resistant clones during treatment can further contribute to therapeutic resistance [136]. Factors secreted by TME, such as cytokines and growth factors, can promote survival pathways, angiogenesis, and immune evasion, contributing to resistance against chemotherapy and targeted therapies [136]. The presence of tumor-associated macrophages (TAMs) or CAFs can modulate signaling pathways and contribute to treatment resistance. CAFs can contribute to therapeutic resistance by secreting factors that remodel the ECM, creating a barrier that limits immune cell recruitment into the tumor [143]. Exosomes, small extracellular vesicles released by cancer cells, can facilitate communication between cancer cells and the TME. These exosomes can transfer molecules, including proteins, lipids, RNA, and DNA, to recipient cells, promoting resistance mechanisms such as enhanced cell survival, angiogenesis, and immune evasion. As treatment is administered, selective pressure is applied to the tumor, leading to subpopulations of cells with altered genetic or phenotypic characteristics that confer resistance to the treatment.

## 2.7 Conclusion

In this chapter, we provide evidence of the involvement of epigenetic changes in BC development and metastasis progression. While driver mutations are commonly found in BC, accumulating studies show epigenetic reprogramming in cancer cells and their surrounding microenvironment. Therefore, it is legitimate to wonder which changes dominate cancer progression, from genetics to epigenetics. Significant areas of interest would be to assess whether (i) the essentiality of genetic mutations is contingent on epigenetic alterations, (ii) epigenetic changes can initiate tumor formation independently of genetic mutations, and (iii) genetic and epigenetic changes are both required for tumor initiation and progression. Current evidence suggests that both work to promote tumor progression, therapeutic resistance, and immune escape. Future investigations may help answer some of these outstanding questions.

## References


1. Servick, K. Breast cancer. Breast cancer: a world of differences. *Science (New York, N.Y.)* **343**, 1452–1453, https://doi.org/10.1126/science.343.6178.1452 (2014).
2. Underwood, J. J. *et al.* Liquid biopsy for cancer: review and implications for the radiologist. *Radiology* **294**, 5–17 (2020).
3. Rizzolo, P., Silvestri, V., Falchetti, M. & Ottini, L. Inherited and acquired alterations in development of breast cancer. *The application of clinical genetics*, 145–158 (2011).
4. Rebbeck, T. R. *et al.* Inheritance of deleterious mutations at both BRCA1 and BRCA2 in an international sample of 32,295 women. *Breast cancer research: BCR* **18**, 112, https://doi.org/10.1186/s13058-016-0768-3 (2016).
5. Immunotherapy, O. C. & Viitala, M. Macrophages to activate antitumor immunity.
6. Krøigård, A. B. *et al.* Identification of metastasis driver genes by massive parallel sequencing of successive steps of breast cancer progression. *PLoS One* **13**, e0189887 (2018).
7. Holsbø, E. & Olsen, K. S. Metastatic breast cancer and pre-diagnostic blood gene expression profiles—the Norwegian Women and Cancer (NOWAC) post-genome cohort. *Frontiers in Oncology* **10**, 575461 (2020).
8. Fasching, P. A. *et al.* Mutations in BRCA1/2 and Other Panel Genes in Patients With Metastatic Breast Cancer-Association With Patient and Disease Characteristics and Effect on Prognosis. *Journal of clinical oncology* **39**, 1619–1630, https://doi.org/10.1200/jco.20.01200 (2021).
9. van Geelen, C. T. *et al.* Clinical implications of prospective genomic profiling of metastatic breast cancer patients. *Breast Cancer Research* **22**, 91, https://doi.org/10.1186/s13058-020-01328-0 (2020).
10. Yazici, H. & Akin, B. Molecular genetics of metastatic breast cancer. *Tumor Progress. Metastasis*, 1–18 (2020).
11. Paul, M. R. *et al.* Genomic landscape of metastatic breast cancer identifies preferentially dysregulated pathways and targets. *The Journal of clinical investigation* **130**, 4252–4265 (2020).
12. Nguyen, B. *et al.* Genomic characterization of metastatic patterns from prospective clinical sequencing of 25,000 patients. *Cell* **185**, 563575, doi:https://doi.org/10.1016/j.cell.2022.01.003 (2022).
13. Karsli-Ceppioglu, S. *et al.* Epigenetic mechanisms of breast cancer: an update of the current knowledge. *Epigenomics* **6**, 651–664 (2014).
14. Consortium, A. P. G. AACR Project GENIE: Powering Precision Medicine through an International Consortium. *Cancer Discov* **7**, 818–831, https://doi.org/10.1158/2159-8290.CD-17-0151 (2017).
15. Thakur, C. *et al.* Epigenetics and environment in breast cancer: New paradigms for anti-cancer therapies. *Front Oncol* **12**, 971288, https://doi.org/10.3389/fonc.2022.971288 (2022).
16. Pouliot, M.-C., Labrie, Y., Diorio, C. & Durocher, F. The role of methylation in breast cancer susceptibility and treatment. *Anticancer research* **35**, 4569–4574 (2015).
17. So, J. Y. *et al.* Induction of DNMT3B by PGE2 and IL6 at Distant Metastatic Sites Promotes Epigenetic Modification and Breast Cancer Colonization. *Cancer Res* **80**, 2612–2627, https://doi.org/10.1158/0008-5472.CAN-19-3339 (2020).
18. An, J., Rao, A. & Ko, M. TET family dioxygenases and DNA demethylation in stem cells and cancers. *Experimental & molecular medicine* **49**, e323–e323 (2017).
19. Wang, W. *et al.* Analysis of methylation-sensitive transcriptome identifies GADD45a as a frequently methylated gene in breast cancer. *Oncogene* **24**, 2705–2714 (2005).
20. Bos, M. K., Smid, M., Sleijfer, S. & Martens, J. W. Apolipoprotein B mRNA-Editing Catalytic Polypeptide-Like–Induced Protein Changes in Estrogen Receptor–Positive, Human Epidermal Growth Factor Receptor 2–Negative Breast Cancer Throughout Disease Progression. *JCO Precision Oncology* **6**, e2100190 (2022).
21. DiMarco, A. V. *et al.* APOBEC mutagenesis inhibits breast cancer growth through induction of T cell–mediated antitumor immune responses. *Cancer immunology research* **10**, 70–86 (2022).
22. Jin, W., Li, Q.-Z., Liu, Y. & Zuo, Y.-C. Effect of the key histone modifications on the expression of genes related to breast cancer. *Genomics* **112**, 853–858 (2020).
23. Zhuang, J., Huo, Q., Yang, F. & Xie, N. Perspectives on the role of histone modification in breast cancer progression and the advanced technological tools to study epigenetic determinants of metastasis. *Frontiers in Genetics* **11**, 603552 (2020).
24. Nam, H. J. & Baek, S. H. Epigenetic Regulation in Breast Cancer. *Advances in experimental medicine and biology* **1187**, 103–119, https://doi.org/10.1007/978-981-32-9620-6_5 (2021).
25. Loh, H.-Y. *et al.* The regulatory role of microRNAs in breast cancer. *International journal of molecular sciences* **20**, 4940 (2019).
26. Si, W., Shen, J., Zheng, H. & Fan, W. The role and mechanisms of action of microRNAs in cancer drug resistance. *Clinical epigenetics* **11**, 1–24 (2019).
27. Ratti, M. *et al.* MicroRNAs (miRNAs) and long non-coding RNAs (lncRNAs) as new tools for cancer therapy: first steps from bench to bedside. *Targeted oncology* **15**, 261–278 (2020).
28. Mollah, S. *Integrative Approach to Targeting Chromatin Remodeling in Breast Cancer Therapy*. Doctoral dissertation. UC San Diego, (2019).

29. Chen, W., Hoffmann, A. D., Liu, H. & Liu, X. Organotropism: new insights into molecular mechanisms of breast cancer metastasis *npj Precision Oncology* **2**, 4, https://doi.org/10.1038/s41698-018-0047-0 (2018).
30. Li, Y. *et al.* The emerging role of ISWI chromatin remodeling complexes in cancer. *Journal of Experimental & Clinical Cancer Research* **40**, 1–27 (2021).
31. Denic, S. & Agarwal, M. M. Breast cancer protection by genomic imprinting in close kin families. *BMC Medical Genetics* **18**, 1–8 (2017).
32. Goovaerts, T. *et al.* A comprehensive overview of genomic imprinting in breast and its deregulation in cancer. *Nature communications* **9**, 4120 (2018).
33. Locke, W. J. *et al.* DNA methylation cancer biomarkers: translation to the clinic. *Frontiers in genetics* **10**, 1150 (2019).
34. Dong, B., Qiu, Z. & Wu, Y. Tackle Epithelial-Mesenchymal Transition With Epigenetic Drugs in Cancer. *Frontiers in pharmacology* **11**, 596239, https://doi.org/10.3389/fphar.2020.596239 (2020).
35. Kong, B. *et al.* Down-regulation of BRMS1 by DNA hypermethylation and its association with metastatic progression in triple-negative breast cancer. *International journal of clinical and experimental pathology* **8**, 11076–11083 (2015).
36. Lim, Y.-Y. *et al.* Epigenetic modulation of the miR-200 family is associated with transition to a breast cancer stem-cell-like state. *Journal of cell science* **126**, 2256–2266 (2013).
37. Cantile, M. *et al.* Long non-coding RNA HOTAIR in breast cancer therapy. *Cancers* **12**, 1197 (2020).
38. Chen, X., Zhang, J. & Dai, X. DNA methylation profiles capturing breast cancer heterogeneity. *BMC Genomics* **20**, 823, https://doi.org/10.1186/s12864-019-6142-y (2019).
39. Bediaga, N. G. *et al.* DNA methylation epigenotypes in breast cancer molecular subtypes. *Breast cancer research* **12**, 1–12 (2010).
40. Holm, K. *et al.* Molecular subtypes of breast cancer are associated with characteristic DNA methylation patterns. *Breast cancer research* **12**, 1–16 (2010).
41. Kim, A. *et al.* Epigenetic Regulation in Breast Cancer: Insights on Epidrugs. *Epigenomes* **7**, https://doi.org/10.3390/epigenomes7010006 (2023).
42. Shargh, S. A. *et al.* Downregulation of E-cadherin expression in breast cancer by promoter hypermethylation and its relation with progression and prognosis of tumor. *Medical oncology* **31**, 1–6 (2014).
43. Mirza, S. *et al.* Promoter hypermethylation of TMS1, BRCA1, ERα and PRB in serum and tumor DNA of invasive ductal breast carcinoma patients. *Life sciences* **81**, 280–287 (2007).
44. Yang, R. *et al.* DNA methylation array analyses identified breast cancer-associated HYAL2 methylation in peripheral blood. *International journal of cancer* **136**, 1845–1855 (2015).
45. Guerrero-Preston, R. *et al.* Differential promoter methylation of kinesin family member 1a in plasma is associated with breast cancer and DNA repair capacity. *Oncology reports* **32**, 505–512 (2014).
46. Heyn, H. *et al.* DNA methylation profiling in breast cancer discordant identical twins identifies DOK7 as novel epigenetic biomarker. *Carcinogenesis* **34**, 102–108 (2013).
47. Kloten, V. *et al.* Promoter hypermethylation of the tumor-suppressor genes ITIH5, DKK3, and RASSF1A as novel biomarkers for blood-based breast cancer screening. *Breast Cancer Research* **15**, 1–11 (2013).
48. Brennan, K. *et al.* Intragenic ATM methylation in peripheral blood DNA as a biomarker of breast cancer risk. *Cancer research* **72**, 2304–2313 (2012).
49. Kuchiba, A. *et al.* Global methylation levels in peripheral blood leukocyte DNA by LUMA and breast cancer: a case–control study in Japanese women. *British journal of cancer* **110**, 2765–2771 (2014).
50. de Almeida, B. P., Apolónio, J. D., Binnie, A. & Castelo-Branco, P. Roadmap of DNA methylation in breast cancer identifies novel prognostic biomarkers. *BMC cancer* **19**, 1–12 (2019).
51. Van Hoesel, A. *et al.* Assessment of DNA methylation status in early stages of breast cancer development. *British journal of cancer* **108**, 2033–2038 (2013).
52. Jovanovic, J., Rønneberg, J. A., Tost, J. & Kristensen, V. The epigenetics of breast cancer. *Molecular oncology* **4**, 242–254 (2010).
53. Park, S. Y. *et al.* Promoter CpG island hypermethylation during breast cancer progression. *Virchows Archiv* **458**, 73–84 (2011).
54. Feng, W. *et al.* Correlation between CpG methylation profiles and hormone receptor status in breast cancers. *Breast cancer research* **9**, 1–13 (2007).
55. Pathiraja, T. N., Stearns, V. & Oesterreich, S. Epigenetic regulation in estrogen receptor positive breast cancer—role in treatment response. *Journal of mammary gland biology and neoplasia* **15**, 35–47 (2010).
56. Ottaviano, Y. L. *et al.* Methylation of the estrogen receptor gene CpG island marks loss of estrogen receptor expression in human breast cancer cells. *Cancer research* **54**, 2552–2555 (1994).
57. Widschwendter, M. *et al.* Association of breast cancer DNA methylation profiles with hormone receptor status and response to tamoxifen. *Cancer research* **64**, 3807–3813 (2004).
58. Lapidus, R. G. *et al.* Mapping of ER gene CpG island methylation by methylation-specific polymerase chain reaction. *Cancer research* **58**, 2515–2519 (1998).
59. Martens, J. W. *et al.* Association of DNA methylation of phosphoserine aminotransferase with response to endocrine therapy in patients with recurrent breast cancer. *Cancer research* **65**, 4101–4117 (2005).

60. Stefansson, O. A. *et al.* CpG island hypermethylation of BRCA1 and loss of pRb as co-occurring events in basal/triple-negative breast cancer. *Epigenetics* **6**, 638–649 (2011).
61. Hsu, N. C. *et al.* Methylation of BRCA1 promoter region is associated with unfavorable prognosis in women with early-stage breast cancer. *PloS one* **8**, e56256 (2013).
62. Khan, S. *et al.* PTEN promoter is methylated in a proportion of invasive breast cancers. *International Journal of Cancer* **112**, 407–410 (2004).
63. García, J. M. *et al.* Promoter methylation of the PTEN gene is a common molecular change in breast cancer. *Genes, Chromosomes and Cancer* **41**, 117–124 (2004).
64. Mehrotra, J. *et al.* Estrogen receptor/progesterone receptor-negative breast cancers of young African-American women have a higher frequency of methylation of multiple genes than those of Caucasian women1. *Clinical cancer research* **10**, 2052–2057 (2004).
65. Dumitrescu, R. G. Epigenetic markers of early tumor development. *Cancer Epigenetics: Methods and Protocols*, 3–14 (2012).
66. Sunami, E. *et al.* Estrogen receptor and HER2/neu status affect epigenetic differences of tumor-related genes in primary breast tumors. *Breast Cancer Research* **10**, 1–11 (2008).
67. Li, L. *et al.* Estrogen and progesterone receptor status affect genome-wide DNA methylation profile in breast cancer. *Human molecular genetics* **19**, 4273–4277 (2010).
68. Jeronimo, C. *et al.* Quantitative hypermethylation of a small panel of genes augments the diagnostic accuracy in fine-needle aspirate washings of breast lesions. *Breast cancer research and treatment* **109**, 27–34 (2008).
69. Radpour, R. *et al.* Methylation profiles of 22 candidate genes in breast cancer using high-throughput MALDI-TOF mass array. *Oncogene* **28**, 2969–2978 (2009).
70. Chimonidou, M. *et al.* CST6 promoter methylation in circulating cell-free DNA of breast cancer patients. *Clinical biochemistry* **46**, 235–240 (2013).
71. Gregory, P. A., Bracken, C. P., Bert, A. G. & Goodall, G. J. MicroRNAs as regulators of epithelial-mesenchymal transition. *Cell cycle* **7**, 3112–3117 (2008).
72. Gregory, P. A. *et al.* The miR-200 family and miR-205 regulate epithelial to mesenchymal transition by targeting ZEB1 and SIP1. *Nature cell biology* **10**, 593–601 (2008).
73. Ye, D., Shen, Z. & Zhou, S. Function of microRNA-145 and mechanisms underlying its role in malignant tumor diagnosis and treatment. *Cancer management and research* **11**, 969–979, https://doi.org/10.2147/cmar.s191696 (2019).
74. Xiao, Y., Humphries, B., Yang, C. & Wang, Z. MiR-205 Dysregulations in Breast Cancer: The Complexity and Opportunities. *Non-coding RNA* **5**, https://doi.org/10.3390/ncrna5040053 (2019).
75. Akhavantabasi, S., Sapmaz, A., Tuna, S. & Erson-Bensan, A. E. miR-125b targets ARID3B in breast cancer cells. *Cell structure and function* **37**, 27–38, https://doi.org/10.1247/csf.11025 (2012).
76. Wang, Y. *et al.* MicroRNA-125b as a tumor suppressor by targeting MMP11 in breast cancer. *Thoracic cancer* **11**, 1613–1620 (2020).
77. Zhai, Z. *et al.* MiR-181a-5p facilitates proliferation, invasion, and glycolysis of breast cancer through NDRG2-mediated activation of PTEN/AKT pathway. *Bioengineered* **13**, 83–95, https://doi.org/10.1080/21655979.2021.2006974 (2022).
78. Zhou, Y. *et al.* The miR-106b~25 cluster promotes bypass of doxorubicin-induced senescence and increase in motility and invasion by targeting the E-cadherin transcriptional activator EP300. *Cell Death & Differentiation* **21**, 462–474, https://doi.org/10.1038/cdd.2013.167 (2014).
79. Powrózek, T. & Małecka-Massalska, T. Is microRNA-31 a key regulator of breast tumorigenesis? *Journal of Thoracic Disease* **10**, 564 (2018).
80. Mayr, C., Hemann, M. T. & Bartel, D. P. Disrupting the pairing between let-7 and Hmga2 enhances oncogenic transformation. *Science (New York, N.Y.)* **315**, 1576–1579 (2007).
81. Yun, J. *et al.* Signalling pathway for RKIP and Let-7 regulates and predicts metastatic breast cancer. *The EMBO Journal* **30**, 4500–4514 (2011).
82. Yu, F. *et al.* let-7 regulates self-renewal and tumorigenicity of breast cancer cells. *Cell* **131**, 1109–1123 (2007).
83. de Souza Rocha Simonini, P. *et al.* Epigenetically deregulated microRNA-375 is involved in a positive feedback loop with estrogen receptor α in breast cancer cells. *Cancer research* **70**, 9175–9184 (2010).
84. Mar-Aguilar, F. *et al.* Serum circulating microRNA profiling for identification of potential breast cancer biomarkers. *Disease markers* **34**, 163–169 (2013).
85. Ma, L., Teruya-Feldstein, J. & Weinberg, R. A. Tumour invasion and metastasis initiated by microRNA-10b in breast cancer. *Nature* **449**, 682–688 (2007).
86. Ma, L. *et al.* Therapeutic silencing of miR-10b inhibits metastasis in a mouse mammary tumor model. *Nature biotechnology* **28**, 341–347 (2010).
87. Bornachea, O. *et al.* EMT and induction of miR-21 mediate metastasis development in Trp53-deficient tumours. *Scientific reports* **2**, 434 (2012).
88. Si, M. *et al.* miR-21-mediated tumor growth. *Oncogene* **26**, 2799–2803 (2007).
89. Si, H. *et al.* Circulating microRNA-92a and microRNA-21 as novel minimally invasive biomarkers for primary breast cancer. *Journal of cancer research and clinical oncology* **139**, 223–229 (2013).
90. Oskarsson, T. *et al.* Breast cancer cells produce tenascin C as a metastatic niche component to colonize the lungs. *Nature medicine* **17**, 867–874 (2011).

91. Zhang, J. *et al.* SOX4 induces epithelial–mesenchymal transition and contributes to breast cancer progression. *Cancer research* **72**, 4597–4608 (2012).
92. Png, K. J. *et al.* MicroRNA-335 inhibits tumor reinitiation and is silenced through genetic and epigenetic mechanisms in human breast cancer. *Genes & development* **25**, 226 (2011).
93. Song, G., Zhang, Y. & Wang, L. MicroRNA-206 targets notch3, activates apoptosis, and inhibits tumor cell migration and focus formation. *Journal of Biological Chemistry* **284**, 31921–31927 (2009).
94. Su, N., Wang, Y., Qian, M. & Deng, M. *2011 IEEE International Conference on Systems Biology (ISB).* 219–224 (IEEE) (20111).
95. Scott, G. K. *et al.* Coordinate suppression of ERBB2 and ERBB3 by enforced expression of micro-RNA miR-125a or miR-125b. *Journal of Biological Chemistry* **282**, 1479–1486 (2007).
96. Wang, S. *et al.* Functional cooperation of miR-125a, miR-125b, and miR-205 in entinostat-induced downregulation of erbB2/erbB3 and apoptosis in breast cancer cells. *Cell death & disease* **4**, e556–e556 (2013).
97. Li, L. *et al.* Targeted expression of miR-34a using the T-VISA system suppresses breast cancer cell growth and invasion. *Molecular Therapy* **20**, 2326–2334 (2012).
98. He, L. *et al.* A microRNA component of the p53 tumour suppressor network. *Nature* **447**, 1130–1134 (2007).
99. Weidhaas, J. B. *et al.* MicroRNAs as potential agents to alter resistance to cytotoxic anticancer therapy. *Cancer research* **67**, 11111–11116 (2007).
100. Neilsen, P. M. *et al.* Mutant p53 drives invasion in breast tumors through up-regulation of miR-155. *Oncogene* **32**, 2992–3000 (2013).
101. Babar, I. A. *et al.* Nanoparticle-based therapy in an in vivo microRNA-155 (miR-155)-dependent mouse model of lymphoma. *Proceedings of the National Academy of Sciences* **109**, E1695–E1704 (2012).
102. Singh, R. & Mo, Y. Y. Role of microRNAs in breast cancer. *Cancer biology & therapy* **14**, 201–212, https://doi.org/10.4161/cbt.23296 (2013).
103. Hussen, B. M. *et al.* MicroRNAs: Important players in breast cancer angiogenesis and therapeutic targets. *Frontiers in Molecular Biosciences* **8**, 764025 (2021).
104. Korkaya, H., Liu, S. & Wicha, M. S. Breast cancer stem cells, cytokine networks, and the tumor microenvironment. *The Journal of clinical investigation* **121**, 3804–3809 (2011).
105. Anderson, N. M. & Simon, M. C. The tumor microenvironment. *Current biology: CB* **30**, R921–R925, https://doi.org/10.1016/j.cub.2020.06.081 (2020).
106. Baghban, R. *et al.* Tumor microenvironment complexity and therapeutic implications at a glance. *Cell Communication and Signaling* **18**, 1–19 (2020).
107. Labani-Motlagh, A., Ashja-Mahdavi, M. & Loskog, A. The tumor microenvironment: a milieu hindering and obstructing antitumor immune responses. *Frontiers in immunology* **11**, 940 (2020).
108. Zhao, H. *et al.* Inflammation and tumor progression: signaling pathways and targeted intervention. *Signal Transduction and Targeted Therapy* **6**, 263, https://doi.org/10.1038/s41392-021-00658-5 (2021).
109. Mun, J.-Y., Leem, S.-H., Lee, J. H. & Kim, H. S. Dual relationship between stromal cells and immune cells in the tumor microenvironment. *Frontiers in Immunology* **13**, 864739 (2022).
110. Lugano, R., Ramachandran, M. & Dimberg, A. Tumor angiogenesis: causes, consequences, challenges and opportunities. *Cellular and molecular life sciences: CMLS* **77**, 1745–1770, https://doi.org/10.1007/s00018-019-03351-7 (2020).
111. Ghalehbandi, S., Yuzugulen, J., Pranjol, M. Z. I. & Pourgholami, M. H. The role of VEGF in cancer-induced angiogenesis and research progress of drugs targeting VEGF. *European Journal of Pharmacology*, 175586 (2023).
112. Mortezaee, K. & Majidpoor, J. Checkpoint inhibitor/interleukin-based combination therapy of cancer. *Cancer Medicine* **11**, 2934–2943 (2022).
113. Arora, S. *et al. Seminars in Cancer Biology* (Elsevier).
114. Cambier, S., Gouwy, M. & Proost, P. The chemokines CXCL8 and CXCL12: Molecular and functional properties, role in disease and efforts towards pharmacological intervention. *Cellular & Molecular Immunology* **20**, 217–251 (2023).
115. Raza, S. *et al. Seminars in cancer biology.* (Elsevier).
116. Pandey, P. *et al.* New insights about the PDGF/PDGFR signaling pathway as a promising target to develop cancer therapeutic strategies. *Biomedicine & Pharmacotherapy* **161**, 114491 (2023).
117. Wang, M. *et al.* Mechanism of immune evasion in breast cancer. *OncoTargets and therapy* **10**, 1561 (2017).
118. Steven, A. & Seliger, B. The role of immune escape and immune cell infiltration in breast cancer. *Breast Care* **13**, 16–21 (2018).
119. Wu, S. Z. *et al.* Stromal cell diversity associated with immune evasion in human triple-negative breast cancer. *The EMBO journal* **39**, e104063 (2020).
120. Rodrigues, J., Heinrich, M. A., Teixeira, L. M. & Prakash, J. 3D in vitro model (R) evolution: unveiling tumor–stroma interactions. *Trends in cancer* **7**, 249–264 (2021).
121. Pei, L. *et al.* Roles of cancer-associated fibroblasts (CAFs) in anti-PD-1/PD-L1 immunotherapy for solid cancers. *Molecular Cancer* **22**, 29 (2023).
122. Hu, D. *et al.* Cancer-associated fibroblasts in breast cancer: Challenges and opportunities. *Cancer Communications* **42**, 401–434 (2022).
123. Mishra, R., Haldar, S., Suchanti, S. & Bhowmick, N. A. Epigenetic changes in fibroblasts drive cancer metabolism and differentiation. *Endocrine-related cancer* **26**, R673–R688 (2019).

124. Dzobo, K. & Dandara, C. The Extracellular Matrix: Its Composition, Function, Remodeling, and Role in Tumorigenesis. *Biomimetics* **8**, 146 (2023).
125. He, X., Lee, B. & Jiang, Y. Extracellular matrix in cancer progression and therapy. *Medical Review* **2**, 125–139 (2022).
126. Yang, L. *et al.* Targeting cancer stem cell pathways for cancer therapy. *Signal transduction and targeted therapy* **5**, 8 (2020).
127. Lüönd, F., Tiede, S. & Christofori, G. Breast cancer as an example of tumour heterogeneity and tumour cell plasticity during malignant progression. *British journal of cancer* **125**, 164–175 (2021).
128. Luo, M., Brooks, M. & MS Wicha, Epithelial-mesenchymal plasticity of breast cancer stem cells: implications for metastasis and therapeutic resistance. *Current pharmaceutical design* **21**, 1301–1310 (2015).
129. Kurma, K. & Alix-Panabières, C. Mechanobiology and survival strategies of circulating tumor cells: a process towards the invasive and metastatic phenotype. *Frontiers in Cell and Developmental Biology* **11**, 1188499 (2023).
130. Rossi, F., Noren, H., Jove, R., Beljanski, V. & Grinnemo, K.-H. Differences and similarities between cancer and somatic stem cells: therapeutic implications. *Stem Cell Research & Therapy* **11**, 1–16 (2020).
131. Yu, Z., Pestell, T. G., Lisanti, M. P. & Pestell, R. G. Cancer stem cells. *The international journal of biochemistry & cell biology* **44**, 2144–2151 (2012).
132. Gkountela, S. *et al.* Circulating Tumor Cell Clustering Shapes DNA Methylation to Enable Metastasis Seeding. *Cell* **176**, 98–112, https://doi.org/10.1016/j.cell.2018.11.046 (2019).
133. Anand, U. *et al.* Cancer chemotherapy and beyond: Current status, drug candidates, associated risks and progress in targeted therapeutics. *Genes & Diseases* (2022).
134. Emran, T. B. *et al.* Multidrug resistance in cancer: understanding molecular mechanisms, immunoprevention and therapeutic approaches. *Frontiers in Oncology* **12**, 2581 (2022).
135. Garcia-Martinez, L., Zhang, Y., Nakata, Y., Chan, H. L. & Morey, L. Epigenetic mechanisms in breast cancer therapy and resistance. *Nature communications* **12**, 1786 (2021).
136. Trédan, O., Galmarini, C. M., Patel, K. & Tannock, I. F. Drug resistance and the solid tumor microenvironment. *Journal of the National Cancer Institute* **99**, 1441–1454 (2007).
137. Du, B. & Shim, J. S. Targeting Epithelial-Mesenchymal Transition (EMT) to Overcome Drug Resistance in Cancer. *Molecules (Basel, Switzerland)* **21**, https://doi.org/10.3390/molecules21070965 (2016).
138. Saha, T. & Lukong, K. E. Breast cancer stem-like cells in drug resistance: a review of mechanisms and novel therapeutic strategies to overcome drug resistance. *Frontiers in Oncology*, 961 (2022).
139. Liu, Y. P. *et al.* Molecular mechanisms of chemo-and radiotherapy resistance and the potential implications for cancer treatment. *MedComm* **2**, 315–340 (2021).
140. Kato, Y. *et al.* Acidic extracellular microenvironment and cancer. *Cancer cell international* **13**, 1–8 (2013).
141. Hudson, K., Cross, N., Jordan-Mahy, N. & Leyland, R. The extrinsic and intrinsic roles of PD-L1 and its receptor PD-1: implications for immunotherapy treatment. *Frontiers in immunology*, 2362 (2020).
142. Wu, M. *et al.* Improvement of the anticancer efficacy of PD-1/PD-L1 blockade via combination therapy and PD-L1 regulation. *Journal of Hematology & Oncology* **15**, 1–58 (2022).
143. Saw, P. E., Chen, J. & Song, E. Targeting CAFs to overcome anticancer therapeutic resistance. *Trends in Cancer* **8**, 527–555 (2022).

# The Epigenetic Landscape of Breast Cancer, Metabolism, and Obesity

3

Ashley Townsel, Maya Jaffe, Yifei Wu, Curtis J. Henry, and Karmella A. Haynes

**Abstract**

Obesity is a risk factor for developing breast cancer, and significantly increases mortality rates in patients diagnosed with this disease. Drivers of this unfortunate relationships are multifactorial, with obesity-induced changes in the epigenetic state of breast cancer cells being identified as a critical mechanism that impact survival, metastasis, and therapeutic responses. Recent studies have investigated the epigenetic landscape of breast cancer to elucidate the molecular interplay between the breast tissue epigenome and its cellular microenvironment. This chapter highlights studies that demonstrates the impact of obesity on the epigenome and metabolome of breast cancer cells. Furthermore, we discuss how obesity impacts the efficacy of chemotherapy and epigenetic targeting drugs, including the emergence of drug-resistance clonal populations. Delineating the relationships between the obesity and epigenetic changes in breast cancer cells will help identify therapeutic strategies which could improve survival outcomes in the rapidly growing number of patients with obesity and cancer.

**Keywords**

Obesity · S-Adenosyl methionine · Intratumoral heterogeneity · T-cell-mediated Immunity · Cancer metabolic pathways

## 3.1 Introduction: Chromatin Dynamics in Breast Cancer

Aberrant epigenetic alterations of chromatin, a nuclear histone protein and DNA complex, deregulate gene expression levels and thus influence the expression levels of proteins associated with cell cycle, apoptosis, and proliferation to promote breast cancer development. The enhancer of zeste homolog (EZH2) methyltransferase, which methylates histone H3 at lysine 27 (H3K27) for example, has been classified as mutated or overexpressed in breast cancer [21, 119]. Hyperactive EZH2 histone methyltransferase activity has been associated with epithelial cell colony formation and invasion in vitro, making its mutational status or overexpression a

A. Townsel
Department of Cancer Biology, Emory School of Medicine, Atlanta, GA, USA

M. Jaffe · Y. Wu · K. A. Haynes (✉)
Wallace H. Coulter Department of Biomedical Engineering, Emory School of Medicine, Atlanta, GA, USA
e-mail: kahayne@emory.edu

C. J. Henry
Department of Immunology and Microbiology, The University of Colorado Anschutz Medical Campus, Aurora, CO, USA


S. Bhatnagar (ed.), *Epigenetic Mechanisms in Breast Cancer Therapy and Resistance*, Advances in Experimental Medicine and Biology 1465, https://doi.org/10.1007/978-3-031-66686-5_3

strong prognostic marker of advanced and metastatic disease [69]. Moreover, mutated epigenetic modifiers UTX and JARID1B/C were demonstrated to have increased de-methyltransferase activity on histone tails at H3K27 and H3K4/K9 positions in breast cancer, respectively [99]. This loss of H3K27 methylation is supported by a study that reveals loss of H3K27me3 among treatment-naive and treatment-resistant TNBC tumors [47]. Overall, patterns of epigenome-wide hypomethylation may serve as a predictive marker of breast cancer risk [122]. Aberrant histone acetylation has also been associated with breast cancer development and can be explored as a therapeutic target [142]. In a study that investigated the epigenetic patterns of different breast cancer subtypes, H3K9ac levels were increased at promoter regions of genes involved with tumorigenesis [65]. The implication of aberrant histone acetylation in basal-like breast cancer is also supported by work that shows decreased tumorigenesis when recognition of histone acetylation marks by bromodomain-containing protein 4 (BRD4) is disrupted [107].

Epigenetic alterations that promote breast cancer progression also involve changes in DNA methylation patterns [16, 50]. In breast cancer and other cancer types, genes that encode tumor-suppressive proteins are commonly found to be repressed via hypermethylation at 5-cytosine-phosphate-guanine-3′ (CpG) regions [33, 59]. In a study that examined DNA methylation patterns across different breast cancer subtypes, DNA methylation was shown to be primarily associated with ER+/luminal breast cancer [49]. These hypermethylation events at tumor-suppressor genes are mediated by mutated or overexpressed wild-type DNA methyltransferase (DNMT), as well as proteins within promoter initiation complexes that maintain DNA methylation in progenitor cells [31, 62, 98, 114]. The genes that are commonly methylated in breast carcinomas include *Cyclin D2*, *RASSF1A*, *BRCA1*, *APC*, and *p16* [35, 41, 117]. These and many other hypermethylated genes found in breast cancer are associated with deregulated DNA repair, genomic and chromosomal instability, aberrant cell motility, anti-apoptosis mechanisms, and aberrant cell cycling [34, 41, 117]. Global methylation and heritable methylation marks are being explored as predictive risk factors for breast cancer [33, 133].

Changes in epigenetic states within cancer progenitor cells can be influenced by external factors including environmental pollutants, high-dose radiation, and adiposity [2, 13, 39]. These factors ultimately increase the risk of breast cancer incidence and can interplay to influence breast cancer progression beyond diagnosis [79]. For instance, women who are obese at the time of breast cancer diagnosis experience poorer survival rates due to tumor-promoting microenvironmental conditions, decreased treatment efficacy, and increased risk of recurrence [73, 132]. While there is strong evidence that links aberrant DNA methylation with breast cancer in general, more work is needed to identify chromatin states that are specifically correlated with breast cancer in patients with obesity [87, 90, 120]. In this chapter, we detail the epigenetic characteristics of breast cancer and review how obesity impacts metabolic states to inform aggressive phenotypes. Epigenetic changes within the broader breast tumor microenvironment are also discussed to highlight the global impact of obesity on breast cancer progression and treatment resistance.

## 3.2 Obesity-Associated Signaling That Influences Metabolism May Also Influence Epigenetic States in Breast Cancer Epithelial Cells

As of 2016, more than 60% of adults worldwide are either overweight or obese—classically defined as having a body mass index (BMI) of >30 [57, 93]. This epidemic is associated with a 52% increased risk of breast cancer onset and decreased survival rates among postmenopausal women diagnosed with breast cancer [25, 72, 88, 91]. In 2021, a regression analysis study was conducted to assess the impact of obesity on breast cancer incidence when preventative treatment is administered [111]. Among the cohort of 3,731 postmenopausal women, a BMI of >30 led to a higher incidence rate of breast cancer. This increased incidence of breast cancer is particularly concerning for Hispanic and non-Hispanic Black women who are not only disproportionately affected by obesity, but also by breast cancer [37, 111]. While these data show high BMI levels to be a risk factor for breast cancer onset, the molecular mechanisms that facilitate breast cancer progression and lead to poor survival rates are not completely understood. This section reviews evidence of how adipocyte-epithelial cell signaling within the tumor microenvironment (TME) affects the levels of epigenetic metabolites within the epithelial cancer cells, and how these changes can affect gene regulation.

### 3.2.1 Secreted Factors from Adipocytes and Obese Sera Affect Metabolic Processes in Breast Epithelial Cells

Obesity is characterized by dysfunctional adipose tissue [19] that secretes excessive fatty acids, hormones, cytokines, and adipokines [3, 25]. Studies have shown that the co-culture of breast cancer cells with obesity-associated adipocytes increases the proliferation and migration of breast cancer cells in vitro [7]. Excessive fatty acids secreted from adipocytes can be directly taken up by breast epithelial cells through the CD36 transmembrane protein, which is overexpressed in drug-resistant breast cancer [36]. In breast cancer cells co-cultured with "obese" adipocytes, increased fatty acid uptake was accompanied by increased fatty acid storage and oxidation, as well as increased conversion of glucose into lipids, and oxidation of glucose and glutamine [7]. Increased glutamine metabolism produces alpha ketoglutarate (α-KG), a key intermediate in the citric acid cycle [80]. Fundamental metabolites, including acetyl-coA, NADH (converted to NAD+ by oxidation), and α-KG, generated by these metabolic pathways serve as key substrates for chromatin remodeling.

Cytokines secreted from adipocytes increase de novo lipid biosynthesis in breast epithelial cells through signal-mediated regulation of fatty acid synthesis genes. Chemokine ligand 5 (CCL5), secreted from adipocytes, binds receptors on breast epithelial cells, and activates the mTOR pathway to regulate cellular growth and metabolism [30]. Downstream of mTOR signaling is the cleavage and activation of the transcriptional factor sterol response element-binding protein-1 (SREBP-1) [71]. Once activated, SREBP-1 translocates to the nucleus to facilitate the transcription of lipogenesis genes, such as fatty acid synthase (FASN) and acetyl-CoA carboxylase (ACC).

In individuals with obesity who consumed a diet that promotes weight gain, high plasma levels of S-adenosylmethionine (SAM) were associated with body mass index and fat mass [32]. The direct effect of this exogenous SAM, originating primarily from liver metabolism, on breast cancer metabolism is not entirely understood. Some experimental evidence suggests that exogenous SAM might support proliferation of breast cancer cells by overcoming the effects of methionine starvation. In vitro studies of methionine-addicted breast cancer cells showed that methionine deprivation led to a decrease in intracellular levels of SAM and reduced their proliferation [15] and that providing these breast cancer cells with exogenous SAM restored their proliferation [14].

In summary, exogenous factors that may be found in an obesity-associated tumor microenvironment, including signaling from obesity-associated adipocytes and excess SAM, can affect metabolic processes in breast epithelial cells. The metabolic processes that are highlighted above involve pools of small metabolites that also serve as substrates for chromatin remodeling enzymes, as discussed in detail in the following sections (Fig. 3.1).

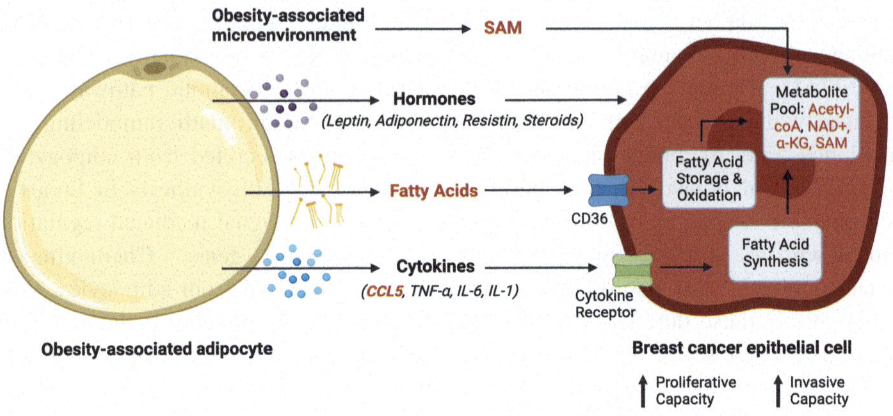


**Fig. 3.1** Obesity-associated adipocytes and the surrounding microenvironment expose breast cancer epithelial cells to factors (hormones, fatty acids, metabolites, and cytokines) that can affect metabolism in the epithelial cells. These environmental signals alter the levels of metabolites involved with chromatin remodeling, including acetyl-coA, NAD+, α-KG, and SAM, to possibly support cancer progression. Recent studies that focus on the roles of extracellular SAM, fatty acids, and the cytokine CCL5 are discussed in detail in this chapter

### 3.2.2 Acetyl-CoA and Acetylation of Histones

The primary substrate for ACC is acetyl-CoA, which catalyzes the formation of malonyl-CoA from acetyl-CoA for fatty acid synthesis [67, 97]. Acetyl-CoA can also be produced via glucose metabolism, utilizing the enzyme ATP-citrate lyase (ACL) [131]. The mechanisms of acetyl-CoA regulation is demonstrated in diabetic murine models that have increased levels of acetyl-CoA [94]. When mice are treated with CCL5, an activator for the mTOR pathway, studies demonstrate that acetyl-CoA levels are reduced, in part, likely due to increased lipogenic consumption of acetyl CoA [44].

Acetyl-CoA is also the sole substrate used by histone acetyltransferase enzymes (HATs) [113] that modify chromatin by adding acetyl groups to histone lysines, which generally promotes transcription and increased gene expression. Therefore, metabolic processes, such as lipogenesis, that affect acetyl-CoA levels are connected to chromatin remodeling and gene regulation [17]. In certain studies, increased acetyl-CoA has been linked to increased histone acetylation in breast and other cancers. In vitro studies with colorectal cancer cells directly connect the rate of glycolysis, which produces acetyl-CoA as a major intermediate, to an increase in histone acetylation levels [27, 40]. This relationship relies on the ACL pathway to use high glucose bioavailability, as is present in obese patients, to increase acetylation levels [131].

Under nutrient-poor conditions, such as a tumor microenvironment, lipogenesis consumes cytosolic acetyl-coA making it unavailable to nuclear HATs [108], which may lead to hypoacetylation at tumor-suppressive genes in cancer cells. In the model organism yeast, investigators showed that blocking the activity of pro-lipogenesis ACC increased levels of global histone lysine acetylation and gene expression [17, 43], presumably by making more acetyl-coA available to HATs. Whether ACC competition for acetyl-coA can lead to decreased histone acetylation in lipogenic breast cancer cells is yet to be determined. So far, studies in breast cancer have shown that a hypoacetylated state and radiation resistance are supported by removal of acetyl groups from histones by overexpressed Sirtuins and histone deacetylase enzymes (HDACs) [106]. This mechanism might synergize with reduced HAT activity to silence anticancer genes. Determining whether obesity-induced lipogenesis contributes to epigenetic silencing of tumor suppressors in breast epithelial cells may inform effective use of HDAC inhibitor treatment in breast cancer patients with obesity [106].

### 3.2.3 S-Adenosylmethionine (SAM) and Methylation of DNA and Histones

Another chromatin remodeling process that is dysregulated in breast cancer, methylation of histones and DNA, is linked to metabolites that are influenced by diet and obesity. Methionine addiction, a dependence on exogenous methionine for proliferation, is a general feature of breast and other cancers and is known as the Hoffman effect (Kaiser [63]). Methionine is converted to S-adenosylmethionine (SAM) through a reaction catalyzed by methionine adenosyltransferases (MATs). When breast cancer cells are deprived of exogenous methionine, intracellular levels of SAM decrease [15]. Supplementing methionine-starved breast cancer cells with exogenous SAM restored their proliferation in vitro, suggesting a cancer-promoting role for extracellular SAM in this context [14]. Exogenous SAM might be supplied systemically (via serum) in individuals with obesity and increased BMI, where liver metabolism is a major producer of circulating SAM [32].

S-adenosylmethionine (SAM) is used as a substrate by DNA methyltransferases (DMNTs) [85] and histone methyltransferases (HMTs) [105] and acts as a methyl donor for other cellular processes [42]. Increased methylation at DNA promoter CpG islands is mediated by DNMTs and is associated with transcriptional repression irrespective of whether the specific target genes have precancerous effects. DNA hypermethylation has, however, been associated with breast cancer metastasis [81]. As for the methylation of histones, EZH2-mediated hypermethylation of H3K27 marks is associated with tumor-suppressor gene repression and drug resistance in breast and other cancers [1, 69]. Therefore, high levels of SAM can enhance cancer-associated transcriptional repression by providing more substrate for DNMTs and HMTs to hypermethylate DNA and histone H3K27.

The molecular impact of excess endogenous SAM on intracellular methylation is a controversial topic however. Contrary to its possible contribution to silencing of tumor suppressors to support cancer cell survival, treatment with SAM showed antiproliferative effects in a xenograft model of the breast cancer cell line MDA-MB-231 [82]. Due to complexities in metabolic flux, excess SAM does not necessarily result in more methylation. A recent study showed that SAM was catabolized to adenine and methylthioadenosine, inhibitors of de novo methylation, in mouse cells in vitro and in vivo [42]. In human liver cancer cells, SAM treatment led to genome-wide hypomethylation as well as hypermethylation of DNA at different loci [128]. More research is needed to determine which subsets of genes in breast cancer cells are silenced or activated during cancer progression in the context of obesity-associated SAM levels and how this regulation contributes to cancer cell proliferation and survival.

### 3.2.4 Alpha-Ketoglutarate and Chromatin Enzymes for Histone and DNA Demethylation

Alpha ketoglutarate (α-KG) is a key intermediate in the citric acid cycle. In cells with high lipogenic activity, upregulated glutamine metabolism potentially increases α-KG levels to support the production of citrate, a key substrate for lipid biosynthesis [80]. Jumonji histone H3K4 demethylases including KDM2A are hyperactive and overexpressed in breast cancer and rely on α-KG as well as 2-oxoglutarate (2OG) [22, 70]. Methylation of histone H3 at lysine 4 (H3K4me) is generally associated with active gene promoters and relatively high gene expression. Similar to histone lysine deacetylation, H4K4 demethylation promotes gene silencing, potentially at tumor suppressors in breast cancer. Therefore, silencing of certain genes by increased α-KG and Jumonji histone H3K4 demethylases might promote breast cancer. The α-KG metabolite is also a substrate for ten-eleven translocase (TET) enzymes, which mediate the removal of methylation at CpG in DNA to allow gene activation (5). In triple-negative breast cancer, TET1 activity promotes the activation of cancer stem cell-promoting gene TAR DNA binding protein 43), which in turn activates serine- and arginine-rich splicing factor 2 (SRSF2) to promote TNBC tumorigenicity [9]. Furthermore, obesity induces a increase in inflammatory signaling, leading to high levels of hydrogen peroxide, which also enhances SRFS2 expression [9, 116]. Contrary to this possible cancer-promoting role, some studies describe an anticancer effect of increased α-KG where accumulation of this metabolite through inhibition of α-ketoglutarate dehydrogenase (KGDH) enhances the levels of nitric oxide (NO) and blocks metastasis to the lungs in a mouse model [5].

### 3.2.5 NAD+ and Chromatin Enzymes PARP and Sirtuins

Nicotinamide adenine dinucleotide (NAD) can exist in two forms, oxidized NAD+ and reduced NADH. NAD+ is a cofactor in reduction-oxidation (redox) reactions associated with energy production, glycolysis, the tricarboxylic acid (TCA) cycle, oxidative phosphorylation, fatty acid oxidation, and serine biosynthesis [86]. The availability of NAD+ for these processes is supported by nicotinamide phosphoribosyltransferase (NAMPT), the key enzyme that recovers NAD from nicotinamide to generate most of the NAD subcellular pools. Circulating NAMPT levels are elevated in obese women and in women suffering from postmenopausal breast cancer [29]. NAMPT may play a protective role for breast cancer cells since high NAMPT expression correlates with poor prognosis and is elevated in chemoresistant breast cancer [130, 139, 143].

Increased NAD+ may also support the activity of chromatin-modifying enzymes that promote survival and proliferation of cancer cells [112]. In response to DNA damage, PARP1 and PARP2 use NAD+ to catalyze ADP-ribosylation of histone serines, which facilitates the recruitment of DNA repair machinery and cancer cell survival [95]. Sirtuins (SIRTs) are class III histone deacetylases and also depend on NAD+ for its activity. SIRT proteins have been associated with deacetylation of histones at, and the epigenetic silencing of, tumor-suppressive apoptosis-promoting genes in breast cancer [96] (Table 3.1).

## 3.3 Single Cell-Level Obesity-Linked Epigenetic States and Therapeutic Resistance

### 3.3.1 Intratumoral Heterogeneity: A Mixed Population of Drug-Tolerant and Drug-Resistant Cells

Chromatin modifications that are affected by adipocyte signaling, such as histone acetylation, histone methylation, and DNA methylation, have been detected at different levels and distributed across the genome differently in epithelial breast cancer cell subpopulations via bulk profiling and in individual cells via single cell 'omics [45, 54, 64]. Epigenetic heterogeneity might influence how individual cells respond to anticancer drugs. The idea of drug resistance via clonal selection [11] proposes that before treatment, the cancer cell population is genetically [18, 67], epigenetically [54], and phenotypically diverse and that treatment eliminates sensitive cells, leaving persisters to restore the cancer cell population.

A study of triple-negative breast cancer revealed that cells within a tumor that had few detectable genomic copy number variations (CNVs) showed highly heterogeneous transcription profiles [64]. To catalog this epigenetic heterogeneity, transcript sequencing has been used to establish subtypes of epithelial cells within tumors (TNBCtype-47): basal-like 1 (BL1), basal-like 2 (BL2), luminal androgen receptor-positive (LAR), and mesenchymal (M). Each subtype's characteristics contribute to pathology and drug response in different ways. In basal-like 1 (BL1) and basal-like 2 (BL2) cells, cell cycle checkpoints are dysregulated and Ki-67 is overexpressed, consistent with rapid proliferation that contributes to tumor growth. The high-cycling nature of these cells renders them sensitive to mitosis-targeting agents such as taxanes and anthracyclines (e.g., doxorubicin). The mesenchymal type (M) includes immunomodulatory

**Table 3.1** Metabolites that are affected by obesity and are used for chromatin remodeling and gene regulation

| Metabolite | Role in obesity-associated metabolism | Role in breast cancer proliferation/invasion | Role in chromatin remodeling in breast cancer cells |
|---|---|---|---|
| Acetyl-coA | Consumed for fatty acid synthesis | Supports fatty acid synthesis and protection from oxidative stress | Used by HATs to acetylate histone lysines |
| SAM | Enhanced hepatic MAT activity in obese individuals uses methionine and ATP to generate SAM. | Can either support or inhibit the proliferation of cells | Used by DNMTs to methylate DNA (CpG); used by HMTs to methylate histones |
| α-KG | Used to produce citrate | Can either support or inhibit the proliferation of cells | Used by Jumonji enzymes to demethylate H3K4; used by TET to remove DNA methylation |
| NAD+ | Reduction-oxidation (redox) reactions for energy production, glycolysis, the tricarboxylic acid (TCA) cycle, oxidative phosphorylation, fatty acid oxidation, and serine biosynthesis | Protection against oxidative stress | Used by PARP1/2 for histone serine ADP-ribosylation; used by Sirtuins to deacetylate histones |

*HAT* histone acetyltransferase, *SAM* S-adenosylmethionine, *MAT* methionine adenosyltransferase, *DNMT* DNA methyltransferase, *HMT* histone methyltransferase, *TET* ten-eleven translocase, *PARP* poly-ADP ribose polymerase, *ATP* adenosine triphosphate, *ADP* adenosine diphosphate, *NAD+* oxidized nicotinamide adenine dinucleotide, *α-KG* alpha ketoglutarate

and stemlike cells. They express genes associated with motility, suggesting a role in metastasis, and are chemoresistant [46]. The luminal androgen receptor-positive (LAR) subtype is the least proliferative and is associated with the lowest pathologic complete response, suggesting strong drug resistance for this subtype [102]. A single tumor can contain all four types of cells, BL1, BL2, LAR, and M17. Therefore, sub-populations of cells (LAR and M) that do not respond to therapy could persist after treatment and lead to cancer relapse.

### 3.3.2 Drug-Resistant Epigenetic States in Migrating Epithelial Cells

Circulating tumor cells (CTC) are cells that migrate from the primary tumor site into the bloodstream to form metastatic lesions in distant sites. The epigenetic changes that influence the metastatic potential of CTCs include DNA methylation and noncoding RNAs (ncRNAs), including microRNA (miRNA) expression [145]. In terms of DNA methylation, hypomethylation of oncogenes and hypermethylation of tumor-suppressor genes promotes tumor progression. Among breast cancer patients, hypermethylation is found at promoters of tumor suppressor genes *CST6* and *SOX17* in CTCs, where *CST6* methylation is associated with EMT [23]. A study demonstrated that the CST6 protein inhibits breast cancer bone metastasis by inhibiting osteoclastogenesis, suggesting hypomethylation of *CST6* could inhibit breast cancer metastasis [78]. Therefore, hypermethylation of *CST6* in CTC could be used as a hallmark to identify breast cancer EMT. For ncRNAs, miRNA-mediated inactivation of tumor-suppressor genes and activation of oncogenes promotes tumor metastasis. For example, microRNA-155 (miR-155) is identified as a hallmark for breast cancer prediction [77]. Wang et al. found that miR-155 overexpression in breast cancer cells is associated with increased T-cell recruitment, tumor growth inhibition, and improved response to immune checkpoint blockade (ICB) therapy [125]. Therefore, higher miR-155 levels in CTCs could be a favorable prognostic marker for breast cancer patients. While DNA hypermethylation in CTCs could be used as biomarkers for breast cancer, the clinical potential of using ncRNAs as epigenetic biomarkers are still under investigation. Noninvasive liquid biopsy techniques are currently a promising avenue for the detection of CTCs, as well as the assessment of breast cancer prognosis and therapy regimens [10]. This technique can help to elucidate the epigenetic landscape of CTCs across metastatic disease states and treatment responses.

## 3.4 Obesity-Induced Epigenetic States in Immune Cells Are Linked to Drug Resistance

### 3.4.1 Myeloid-Derived Suppressor Cells

Myeloid-derived suppressor cells (MDSC) are a heterogeneous population of pathologically activated immature myeloid cells with potent tumor-favoring immunosuppressive effect [121]. MDSCs are found to impair the efficacy of multiple cancer treatments, such as radiotherapy, chemotherapy, and immunotherapy, and therefore have been identified as a therapeutic target for antitumor therapies [75]. In an obese microenvironment, MDSC levels are elevated and obesity-induced MDSCs promote tumor progression [26]. Epigenetic mechanisms, such as DNA methylation, play an essential role in regulating how MDSCs behave in an obese TME. Δ9-Tetrahydrocannabinol (THC), an exogenous cannabinoid derived from the *Cannabis Sativa* plant, is found to induce decreased methylation at the promoter region of *STAT3* in THC-induced MDSCs, resulting in increased expression of STAT3 [110]. Elevated levels of STAT3 protein levels are found in breast cancer-derived MDSCs and STAT3-miRNA allows the inhibition of immunosuppressive effects of MDSCs [100]. Therefore, THC could be used as a novel antitumor therapy to target STAT3 expression in obesity-induced MDSCs under the breast cancer TME to alleviate chemoresistance, which is found to be induced through the EZH2/STAT3 pathway in breast cancer cells [136].

### 3.4.2 M1/M2 Macrophages and Tumor-Associated Macrophages

Obesity is characterized by chronic macrophage-induced inflammation. Macrophages are divided into M1-like and M2-like macrophages. M1 macrophages mainly produce pro-inflammatory cytokines, while M2 macrophages mainly produce anti-inflammatory cytokines [137]. Tumor-associated macrophages (TAM) forming crown-like structures surrounding the necrotic adipocytes is considered as a hallmark of obesity. TAM can contribute to tumor cell proliferation, invasion, cancer stem cell-like behavior, and chemoresistance by secreting tumor-promoting cytokines. Under obesity, TAM can contribute to the deposition of extracellular matrix (ECM), leading to increased tumor recurrence and metastasis through ECM remodeling [53]. In an obese microenvironment, DNA methylation by DNA methyltransferase 1 (DNMT1) regulates TAM inflammation and polarization on the epigenetic level. For instance, obesity-associated factors, such as dietary saturated fatty acids (SFAs) and pro-inflammatory cytokines, can significantly enhance DNMT1-mediated *PPARγ1* promoter DNA methylation in macrophages, resulting in decreased adipose tissue macrophages (ATM) alternative activation, increased ATM inflammation, and decreased insulin sensitivity [126]. This study provides guidance on the development of new epidrugs that could regulate DNA methylation in TAM in the breast cancer TME. Histone methyltransferase enhancer of Zeste homolog 2 (EZH2) has been shown to interact with histone H3 at lysine 27 (H3K27) to impact tumor metastasis via regulation of macrophage polarization. For instance, EZH2 inhibitors (EZH2i) inhibit trimethylated lysine 27 of histone H3 (H3K27me3) levels on the promoter of *STAT3*, leading to decreased M-2 macrophage proportion and increased M-1 macrophage in the TME. By skewing M-2 macrophage into M-1 macrophage, EZH2 inhibition achieves a tumor-suppressive effect [76]. Under an obese microenvironment, EZH2 also mediates H3K27me3 to regulate lipogenesis. A study showed EZH2 methylates

H3K27 to promote white adipocytes while inhibiting white and beige adipocytes differentiation, suggesting H3K27me3 as a key epigenetic switch to determine adipocyte differentiation [134]. In addition, EZH2 has been identified as a target to alleviate endocrine resistance. For instance, the bone microenvironment has been demonstrated to enhance the phenotypic plasticity of $ER^+$ breast cancer via EZH2-mediated epigenetic programming, leading to increased stemness and endocrine resistance. Inhibition of EZH2 restores endocrine sensitivity and could be used as a novel tool in breast cancer therapeutic intervention [6].

### 3.4.3 Obesity and T-Cell-Mediated Immunity

One hallmark of obesity is loss of immune homeostasis. Nearly every immune cell is impacted by increased adiposity, which commonly results in attenuated immune function. This is particularly evident in the case of T-cell-mediated immunity, where CD4+ T helper cells and CD8+ cytotoxic T cells are documented to have homing, proliferative, cytokine-producing, and cytotoxic defects [138].

### 3.4.4 T-Cell Homing Alterations

Obesity induces physiological changes in hosts including the induction of systemic low-grade chronic inflammation, which can impact the homing of immune cells. It has been documented that local and systemic changes in cytokines and chemokines promote the recruitment of immune cells to adipose tissue and away from the lymphatic system [68, 83]. The increased homing of CD4+ and CD8+ T cells among adipocytes is due, in part, to changes in the immunophenotype (surface expression of cytokine and chemokine receptors) of effector T cells [68, 83]. For example, when CD4+ and CD8+ T cells are activated in mice-fed high-fat diets, a higher frequency of this population upregulate their surface expression of CD44 (an adhesion molecule which binds to extracellular matrix proteins; [84]), CXCR3 (a chemokine receptor which recognizes the pro-inflammatory chemokines CXCL9, CXCL10, and CXCL11; [127]), and LFA-1 (an integrin which binds ICAM-1, ICAM-2, and ICAM-3; [12]). Concurrently, these cells also downregulate a chemokine receptor, CCR7, which recognizes the chemokines CCL19 and CCL21 produced by lymphatic cells [38], and a lectin, CD62L, which binds to a multitude of lectins with a high specificity for those with sialyl Lewis X [sLex] carbohydrate structures [60]. In addition to the heightened attraction of effector T cells to adipose tissue resulting from obesity [68], once T cell localize to these sites, they upregulate their surface expression of killer-cell leptin-like receptor G1 (KLRG1) which attenuates their cytolytic capacity [52]. Overall, obesity-induced changes in the immunophenotype of T cells increase their localization to adipose tissue, which in turn diminishes their function. The local and systemic impact of increased adiposity on T-cell function will be discussed in greater detail below.

### 3.4.5 T-Cell Functional Changes

The impact of obesity in attenuating T-cell proliferation, cytokine production, and cytolysis is well documented [141]; however, we are just beginning to delineate the mechanisms of adipocyte-mediated T-cell immunosuppression. Naïve T cells are activated by 3 signals: (1) ligation to cognate antigen typically when the T-cell receptor (TCR) recognizes peptides presented by major histocompatibility complex (MHC) proteins, (2) costimulation via costimulatory molecules to propagate TCR-mediated signaling, and (3) polarization of the response directed by cytokine signaling [103]. Given the pro-inflammatory nature of adipose-rich microenvironments [66], soluble changes in local and systemic metabolites [140], and adipocyte-induced changes in tissue architecture [4], among others, the impact of adiposity on T-cell activation and function is multifactorial and ultimately results in T-cell exhaustion [92, 109]. Recently, a study of adipose tissue T cells from mice and humans demonstrated that adipocyte-induced T-cell exhaustion resulted from exposure to soluble factors from the stromal vascular fraction and macrophage-mediated immune suppression [92]. Notably, regarding the latter, macrophages suppressed T-cell function in adipose tissue in an MHC class II-dependent, PD-1 independent manner. Despite not observing PD-1-mediated immunosuppression in this microenvironment, a greater frequency of adipose tissue T cells upregulated the co-inhibitory receptor B- and T-lymphocyte attenuator (BTLA). This inhibitory protein contains src homology region 2 domain-containing phosphatase 1 (SHP-1) and src homology region 2 domain-containing phosphatase 2 (SHP-2) on its cytoplasmic tail, which would attenuate TCR signaling [92].

In addition to obesity-mediated impairments of TCR signaling, systemic leptin levels are higher due to obesity and induces T-cell exhaustion through heightened STAT3 signaling [8, 129]. Interleukin-6 (IL-6) is an adipokine which is found at elevated local and systemic levels due to obesity and this cytokine is also a potent activator of STAT3 signaling in T cells [92]. Therefore, obesity-induced T-cell exhaustion is also propagated by heightened STAT3 signaling resulting from higher leptin and IL-6 levels in adipocyte-rich microenvironments.

### 3.4.6 Epigenetic Modifications and Considerations for Immunotherapies

An elegant study conducted on myeloid cells has demonstrated that adipose tissue myeloid cells (ATMs) exposed to a high-fat diet and switched to a normal chow diet maintain epigenetic signatures indicative of mice maintained exclusively on high-fat diets. Specifically, the authors found that stearic acid stimulation of Toll-like receptor 4 (TLR4) remodels the chromatin landscape and enhances accessibility to binding sites for activator protein-1 (AP-1). Using transposase-accessible chromatin sequencing (ATAC-seq), the authors demonstrate that switching mice from high-fat diet to normal chow after 11 weeks still resulted in ATMs with a more open chromatin conformation similar to ATMs isolated from mice fed exclusively with high-fat diets. Analysis of differential accessibility regions (DARs), using gene ontology (GO) enrichment analysis, revealed that ATMs isolated from mice originally fed high-fat diet and switched to normal chow resulted in the hyperactivation of pathways in ATMs associated with angiogenesis, cytokine productions, and inflammatory responses. Given the long-term impact of obesity on the epigenome of myeloid cells, it stands to reason that similar changes will be induced in T cells due to adiposity. Developing a deeper understanding of obesity-induced changes in the epigenome of T cells will be instrumental for optimizing immunotherapies targeting solid and hematological malignancies including breast cancer [51].

Obesity significantly decreases the survival of patients with breast cancer [73] and, as indicated above, is associated with epigenetic reprogramming of myeloid cells which are not easily reversible with dietary interventions alone. This may also be the case for T cells, and both immune cell populations are large components of the breast cancer tumor microenvironment. This observation should be taken into consideration in patients receiving the recently FDA-approved immune checkpoint inhibitors (ICI) for triple-negative breast cancer (TNBC), atezolizumab and pembrolizumab, which target the PD-L1/2 and PD-1/2 pathway. The importance of continued scientific investments into understanding how epigenetics could impact the efficacy of immunotherapies will be particularly important given the observation that exhausted T cells are present at high percentages in solid cancer including breast cancer [61, 118]. In studies of chronic viral infections, it was shown that the epigenetic profile of exhausted T cells differ substantially from those of effector and memory T cells [89, 104]. Interestingly, an epigenetic examination of exhausted T cells after PD-1 blockade treatment revealed that transcriptional reprogramming occurs; however, T cells in the tumor microenvironment never acquired a memory T-cell epigenetic imprint and reverted to an exhausted phenotype if antigen density remains high [89].

## 3.5 Conclusions: Implications for Epigenetic Therapy

The growing body of evidence discussed in this chapter suggests a mechanistic link between obesity and changes in epigenetic states in breast cancer cells that lead to poor treatment outcomes. Pools of key metabolites that are used for chromatin remodeling in breast epithelial cells, as well as other cells in the tumor microenvironment, are affected by obesity through adipocyte-secreted factors and molecules that are elevated in the sera of individuals with obesity. Therefore, breast cancer patients with obesity might benefit from epigenetic therapies that target obesity-associated epigenetic states. In the following section, we discuss epigenetic drugs that target chromatin enzymes of which use obesity-affected metabolites as substrates.

In the context of obesity, histone lysine hypoacetylation may be supported by the depletion of acetyl-coA through lipogenesis and enhanced removal of acetyl groups from histone lysines by HDACs and Sirtuins. Enhanced hypoacetylation by Sirtuins may also be supported by increased availability of NAD+. While HDAC inhibition has shown promise in the in vitro setting, the successful use of HDAC inhibitors have not yet translated into the clinical treatment of breast cancer patients [56, 74, 144]. In preclinical murine models, HDAC inhibitors were actually shown to promote metastasis [58]. Also in an obese context, increased levels of SAM might support increased activity of histone H2K27 methylation and DNA methylation, which are mediated by EZH2 and DNMTs, respectively. The epigenetic trends that are characteristic of breast cancer involve deregulated histone acetylation via mutated or upregulated EZH2 activity [21, 119]. The imbalances in these histone modifying enzymes contribute to epigenetic landscapes that promote mechanisms of breast cancer stemness and oncogenesis. Tazemetostat is an FDA-approved EZH2 methyltransferase inhibitor for patients with follicular lymphoma and epithelioid lymphoma [55] but has not demonstrated tumorigenic inhibition of breast cancer cells in the clinic. While the promise of combining tazemetostat with other drugs is being explored, groups are developing drug analogs to improve its efficacy in breast cancer [101]. In a study that developed and assessed EZH2 inhibitors, their U3i compound was found to be more efficacious in selectively reducing TNBC growth compared to tazemetostat by more than 70-fold [123]. Regulators of EZH2 expression could also be likely targets for EZH2 inhibition, but off-target effects would need to be studied more closely. For example, HIF1 is a transcription factor that gets upregulated in breast tumor initiating cells in hypoxic conditions and thereby upregulates EZH2 expression [20]. Moreover, knockdown of O-linked N-acetylglucosamine (GlcNAc) transferase (OGT) was shown to selectively decrease EZH2 stability, making it a potential therapeutic target [24].

NAD+ availability is affected by changes in obesity-associated fatty acid metabolism, is enhanced by elevated levels of NAMPT in obese individuals, and is a key metabolite used by PARP 1 and 2. PARP1/2 modify chromatin by ADP-ribosylation of histone serines, which protects cancer cells from the effects of DNA breaks. The PARP inhibitors, olaparib (Lynparza) and talazoparib (Talzenna), have been FDA approved for treatment of HER2-negative metastatic breast cancer [28]. Considering PARP proteins interact with EZH2, the effect of PARP inhibition on the activity of EZH2 methyltransferase activity was investigated in TNBC cells. Treatment of TNBC cells with olaparib demonstrated increased global H3K27me3 levels [135] which supports gene repression patterns.

Combination therapies using the epigenetic drugs discussed here, as well as other anticancer agents, may be a promising approach for treating breast cancer patients with obesity. Newly designed dual inhibitors of both EZH2 and PARP have demonstrated the ability to sensitize BRCA wild-type breast cancer cells to PARP inhibitor drugs [124, 135]. Dual target inhibitors of histone methyltransferase EZH2 and histone acetylase BRD4 have also been developed and demonstrated to decrease breast cancer growth and proliferation compared to tazemetostat (EZH2i) alone [48]. Clinical trials investigating the efficacy of combining HDAC inhibitors (HDACi) with immunotherapy drugs are preceded by preclinical murine models of TNBC that demonstrate increases in immunomodulatory protein expression with vorinostat (HDACi) treatment [115]. Treatment of metastatic breast cancer with entinostat (HDACi) and nivolumab (anti-CTLA4) combination therapy with and without ipilimumab (anti-CTLA4) is being investigated in a current phase 1 clinical trial (NCT02453620).

Mechanistic models that emerge from experimental evidence suggest that breast cancer patients with obesity, compared to lean patients, might respond differently to epigenetic drugs that target obesity-affected chromatin states, but more work is needed to determine a clear benefit. For instance, future clinical studies should compare treatment efficacy in breast cancer patients with or without obesity to clearly determine the clinical relevance of the preclinical data. Furthermore, preclinical epigenetic drug studies tend to be done in the absence of obesity-associated adipocytes and their secreted factors. Filling these gaps could overcome current barriers toward a cure for breast cancers in patients with diverse body types.

## References


1. Adibfar S, Elveny M, Kashikova HS, Mikhailova MV, Farhangnia P, Vakili-Samiani S, et al. The molecular mechanisms and therapeutic potential of EZH2 in breast cancer. Life Sci. 2021 Dec 1;286:120047.
2. Alegría-Torres JA, Baccarelli A, Bollati V. Epigenetics and lifestyle. Epigenomics 2011 Jun;3(3):267–77.
3. Al-Mansoori L, Al-Jaber H, Prince MS, Elrayess MA. Role of Inflammatory Cytokines, Growth Factors and Adipokines in Adipogenesis and Insulin Resistance. Inflammation. 2022 Feb;45(1):31–44.
4. Asghar A, Akhtar T, Batool T, Khawar MB, Nadeem S, Mehmood R, et al. High-fat diet-induced splenic, hepatic, and skeletal muscle architecture damage: cellular and molecular players. Mol. Cell. Biochem. 2021 Oct;476(10):3671–9.
5. Atlante S, Visintin A, Marini E, Savoia M, Dianzani C, Giorgis M, et al. α-ketoglutarate dehydrogenase inhibition counteracts breast cancer-associated lung metastasis. Cell Death Dis. 2018 Jul 9;9(7):756.
6. Bado IL, Zhang W, Hu J, Xu Z, Wang H, Sarkar P, et al. The bone microenvironment increases phenotypic plasticity of ER breast cancer cells. Dev. Cell. 2021 Apr 19;56(8):1100–17.e9.
7. Balaban S, Shearer RF, Lee LS, van Geldermalsen M, Schreuder M, Shtein HC, et al. Adipocyte lipolysis links obesity to breast cancer growth: adipocyte-derived fatty acids drive breast cancer cell proliferation and migration. Cancer Metab. 2017 Jan 13;13;5:1.
8. Bally APR, Austin JW, Boss JM. Genetic and Epigenetic Regulation of PD-1 Expression. J. Immunol. 2016 Mar 15;196(6):2431–7.
9. Bao B, Teslow EA, Mitrea C, Boerner JL, Dyson G, Bollig-Fischer A. Role of TET1 and 5hmC in an Obesity-Linked Pathway Driving Cancer Stem Cells in Triple-Negative Breast Cancer. Mol. Cancer Res. 2020 Dec;18(12):1803–14.
10. Bao-Caamano A, Rodriguez-Casanova A, Diaz-Lagares A. Epigenetics of Circulating Tumor Cells in Breast Cancer. Adv. Exp. Med. Biol. 2020;1220:117–34.
11. Bedard PL, Hansen AR, Ratain MJ, Siu LL. Tumour heterogeneity in the clinic. Nature. 2013 Sep 19;501(7467):355–64.
12. Binnerts ME, van Kooyk Y. How LFA-1 binds to different ligands. Immunol. Today. 1999 May;20(5):240–5.
13. Bollati V, Baccarelli A. Environmental epigenetics. Heredity. 2010 Jul;105(1):105–12.
14. Booher K, Lin D-W, Borrego SL, Kaiser P. Downregulation of Cdc6 and pre-replication complexes in response to methionine stress in breast cancer cells. Cell Cycle. 2012 Dec 1;11(23):4414–23.
15. Borrego SL, Fahrmann J, Datta R, Stringari C, Grapov D, Zeller M, et al. Metabolic changes associated with methionine stress sensitivity in MDA-MB-468 breast cancer cells. Cancer Metab. 2016 May 2;4:9.
16. Buitrago D, Labrador M, Arcon JP, Lema R, Flores O, Esteve-Codina A, et al. Impact of DNA methylation on 3D genome structure. Nat. Commun. 2021 May 28;12(1):3243.
17. Cai L, Sutter BM, Li B, Tu BP. Acetyl-CoA induces cell growth and proliferation by promoting the acetylation of histones at growth genes. Mol. Cell. 2011 May 20;42(4):426–37.
18. Casasent AK, Schalck A, Gao R, Sei E, Long A, Pangburn W, et al. Multiclonal Invasion in Breast Tumors Identified by Topographic Single Cell Sequencing. Cell. 2018 Jan 11;172(1–2):205–17.e12.
19. Chait A, den Hartigh LJ. Adipose Tissue Distribution, Inflammation and Its Metabolic Consequences, Including Diabetes and Cardiovascular Disease. Front Cardiovasc Med. 2020 Feb 25;7:22.
20. Chang C-J, Yang J-Y, Xia W, Chen C-T, Xie X, Chao C-H, et al. EZH2 promotes expansion of breast tumor initiating cells through activation of RAF1-β-catenin signaling. Cancer Cell. 2011 Jan 18;19(1):86–100.

21. Chase A, Cross NCP. Aberrations of EZH2 in cancer. Clin. Cancer Res. 2011 May 1;17(9):2613–8.
22. Chen J-Y, Luo C-W, Lai Y-S, Wu C-C, Hung W-C. Lysine demethylase KDM2A inhibits TET2 to promote DNA methylation and silencing of tumor suppressor genes in breast cancer. Oncogenesis. 2017 Aug 7;6(8):e369.
23. Chimonidou M, Strati A, Tzitzira A, Sotiropoulou G, Malamos N, Georgoulias V, et al. DNA methylation of tumor suppressor and metastasis suppressor genes in circulating tumor cells. Clin. Chem. 2011 Aug;57(8):1169–77.
24. Chu C-S, Lo P-W, Yeh Y-H, Hsu P-H, Peng S-H, Teng Y-C, et al. O-GlcNAcylation regulates EZH2 protein stability and function. Proc. Natl. Acad. Sci. U. S. A. 2014 Jan 28;111(4):1355–60.
25. Chu D-T, Phuong TNT, Tien NLB, Tran D-K, Nguyen T-T, Van Thanh V, et al. The Effects of Adipocytes on the Regulation of Breast Cancer in the Tumor Microenvironment: An Update. Cells 2019 Aug 8;8(8). https://doi.org/10.3390/cells8080857
26. Clements VK, Long T, Long R, Figley C, Smith DMC, Ostrand-Rosenberg S. Frontline Science: High fat diet andleptin promote tumor progression by inducing myeloid-derived suppressor cells. J Leukoc Biol. 2018 Mar 103(3):395–407. https://doi.org/10.1002/JLB.4HI0517-210R.
27. Cluntun AA, Huang H, Dai L, Liu X, Zhao Y, Locasale JW. The rate of glycolysis quantitatively mediates specific histone acetylation sites. Cancer Metab. 2015 Spe 23;3:10.
28. Cortesi L, Rugo HS, Jackisch C. An Overview of PARP Inhibitors for the Treatment of Breast Cancer. Target. Oncol. 2021 May;16(3):255–82.
29. Dalamaga M. Nicotinamide phosphoribosyltransferase/visfatin: a missing link between overweight/obesity and postmenopausal breast cancer? Potential preventive and therapeutic perspectives and challenges. Med. Hypotheses. 2012 Nov;79(5):617–21.
30. D'Esposito V, Liguoro D, Ambrosio MR, Collina F, Cantile M, Spinelli R, et al. Adipose microenvironment promotes triple negative breast cancer cell invasiveness and dissemination by producing CCL5. Oncotarget. 2016 Apr 26;7(17):24495–509.
31. Dunican DS, Cruickshanks HA, Suzuki M, Semple CA, Davey T, Arceci RJ, et al. Lsh regulates LTR retrotransposon repression independently of Dnmt3b function. Genome Biol. BioMed Central; 2013 Dec 24;14(12):1–23.
32. Elshorbagy AK, Jernerén F, Samocha-Bonet D, Refsum H, Heilbronn LK. Serum S-adenosylmethionine, but not methionine, increases in response to overfeeding in humans. Nutr. Diabetes. 2016 Jan 25;6(1):e192.
33. Ennour-Idrissi K, Dragic D, Durocher F, Diorio C. Epigenome-wide DNA methylation and risk of breast cancer: a systematic review. BMC Cancer. 2020 Oct 31;20(1):1048.
34. Esteller M. Epigenetic lesions causing genetic lesions in human cancer: promoter hypermethylation of DNA repair genes. Eur. J. Cancer. 2000 Dec;36(18):2294–300.
35. Esteller M, Silva JM, Dominguez G, Bonilla F, Matias-Guiu X, Lerma E, et al. Promoter hypermethylation and BRCA1 inactivation in sporadic breast and ovarian tumors. J. Natl. Cancer Inst. 2000 Apr 5;92(7):564–9.
36. Feng WW, Wilkins O, Bang S, Ung M, Li J, An J, et al. CD36-Mediated Metabolic Rewiring of Breast Cancer Cells Promotes Resistance to HER2-Targeted Therapies. Cell Rep. 2019 Dec10;29(11):3405–20.e5.
37. Flegal KM, Carroll MD, Kit BK, Ogden CL. Prevalence of obesity and trends in the distribution of body mass index among US adults, 1999-2010. JAMA. 2012 Feb 1;307(5):491–7.
38. Förster R, Davalos-Misslitz AC, Rot A. CCR7 and its ligands: balancing immunity and tolerance. Nat. Rev. Immunol. 2008 May;8(5):362–71.
39. Fraga MF. Genetic and epigenetic regulation of aging. Curr. Opin. Immunol. 2009 Aug;21(4):446–53.
40. Friis RMN, Wu BP, Reinke SN, Hockman DJ, Sykes BD, Schultz MC. A glycolytic burst drives glucose induction of global histone acetylation by picNuA4 and SAGA. Nucleic Acids Res. 2009 Jul;37(12):3969–80.
41. Fujikane T, Nishikawa N, Toyota M, Suzuki H, Nojima M, Maruyama R, et al. Genomic screening for genes upregulated by demethylation revealed novel targets of epigenetic silencing in breast cancer. Breast Cancer Res. Treat. 2010 Aug;122(3):699–710.
42. Fukumoto K, Ito K, Saer B, Taylor G, Ye S, Yamano M, et al. Excess S-adenosylmethionine inhibits methylation via catabolism to adenine. Commun Biol. 2022 Apr 5;5(1):313.
43. Galdieri L, Vancura A. Acetyl-CoA carboxylase regulates global histone acetylation. J. Biol. Chem. 2012 Jul 6;287(28):23865–76.
44. Gao D, Rahbar R, Fish EN. CCL5 activation of CCR5 regulates cell metabolism to enhance proliferation of breast cancer cells. Open Biol. 2016 Jun;6(6). https://doi.org/10.1098/rsob.160122
45. Gao R, Kim C, Sei E, Foukakis T, Crosetto N, Chan L-K, et al. Nanogrid single-nucleus RNA sequencing reveals phenotypic diversity in breast cancer. Nat. Commun. 2017 Aug 9;8(1):228.
46. Gibson GR, Qian D, Ku JK, Lai LL. Metaplastic breast cancer: clinical features and outcomes. Am. Surg. 2005 Sep;71(9):725–30.
47. Grosselin K, Durand A, Marsolier J, Poitou A, Marangoni E, Nemati F, et al. High-throughput single-cell ChIP-seq identifies heterogeneity of chromatin states in breast cancer. Nat. Genet. 2019 Jun;51(6):1060–6.
48. Guo Z, Sun Y, Liang L, Lu W, Luo B, Wu Z, et al. Design and Synthesis of Dual EZH2/BRD4 Inhibitors to Target Solid Tumors. J. Med. Chem. 2022 May 12;65(9):6573–92.

49. Győrffy B, Bottai G, Fleischer T, Munkácsy G, Budczies J, Paladini L, et al. Aberrant DNA methylation impacts gene expression and prognosis in breast cancer subtypes. Int. J. Cancer. 2016 Jan 1;138(1):87–97.
50. Hashimshony T, Zhang J, Keshet I, Bustin M, Cedar H. The role of DNA methylation in setting up chromatin structure during development. Nat. Genet. 2003 Jun;34(2):187–92.
51. Hata M, Andriessen EMMA, Hata M, Diaz-Marin R, Fournier F, Crespo-Garcia S, et al. Past history of obesity triggers persistent epigenetic changes in innate immunity and exacerbates neuroinflammation. Science. 2023 Jan 6;379(6627):45–62.
52. Henson SM, Akbar AN. KLRG1--more than a marker for T cell senescence. Age. 2009 Dec;31(4):285–91.
53. Hillers-Ziemer LE, Kuziel G, Williams AE, Moore BN, Arendt LM. Breast cancer microenvironment and obesity: challenges for therapy. Cancer Metastasis Rev. 2022 Sep;41(3):627–47.
54. Hinohara K, Polyak K. Intratumoral Heterogeneity: More Than Just Mutations. Trends Cell Biol. 2019 Jul;29(7):569–79.
55. Hoy SM. Tazemetostat: First Approval. Drugs. 2020 Apr;80(5):513–21.
56. Huang M, Zhang J, Yan C, Li X, Zhang J, Ling R. Small molecule HDAC inhibitors: Promising agents for breast cancer treatment. Bioorg. Chem. 2019 Oct;91:103184.
57. Hurt RT, Kulisek C, Buchanan LA, McClave SA. The obesity epidemic: challenges, health initiatives, and implications for gastroenterologists. Gastroenterol. Hepatol. 2010 Dec;6(12):780–92.
58. Hu Z, Wei F, Su Y, Wang Y, Shen Y, Fang Y, et al. Histone deacetylase inhibitors promote breast cancer metastasis by elevating NEDD9 expression. Signal Transduction and Targeted Therapy. Nature Publishing Group; 2023 Jan 6;8(1):1–13.
59. Ibrahim J, Op de Beeck K, Fransen E, Peeters M, Van Camp G. Genome-wide DNA methylation profiling and identification of potential pan-cancer and tumor-specific biomarkers. Mol. Oncol. 2022 Jun;16(12):2432–47.
60. Ivetic A, Hoskins Green HL, Hart SJ. L-selectin: A Major Regulator of Leukocyte Adhesion, Migration and Signaling. Front. Immunol. 2019 May 14;14;10:1068.
61. Jiang W, He Y, He W, Wu G, Zhou X, Sheng Q, et al. Exhausted CD8+T Cells in the Tumor Immune Microenvironment: New Pathways to Therapy. Front. Immunol. 2020;11:622509.
62. Joo JE, Dowty JG, Milne RL, Wong EM, Dugué P-A, English D, et al. Heritable DNA methylation marks associated with susceptibility to breast cancer. Nat. Commun. 2018 Feb 28;9(1):867.
63. Kaiser P. Methionine Dependence of Cancer. Biomolecules 2020 Apr 8;10(4). https://doi.org/10.3390/biom10040568
64. Karaayvaz M, Cristea S, Gillespie SM, Patel AP, Mylvaganam R, Luo CC, et al. Unravelling subclonal heterogeneity and aggressive disease states in TNBC through single-cell RNA-seq. Nat. Commun. 2018 Sep 4; 9(1):3588.
65. Karsli-Ceppioglu S, Dagdemir A, Judes G, Lebert A, Penault-Llorca F, Bignon Y-J, et al. The Epigenetic Landscape of Promoter Genome-wide Analysis in Breast Cancer. Sci. Rep. 2017 Jul 26;7(1):6597.
66. Kawai T, Autieri MV, Scalia R. Adipose tissue inflammation and metabolic dysfunction in obesity. Am. J. Physiol. Cell Physiol. 2021 Mar 1;320(3):C375–91.
67. Kim C, Gao R, Sei E, Brandt R, Hartman J, Hatschek T, et al. Chemoresistance Evolution in Triple-Negative Breast Cancer Delineated by Single-Cell Sequencing. Cell. 2018 May 3;173(4):879–93.e13.
68. Kiran S, Kumar V, Murphy EA, Enos RT, Singh UP. High Fat Diet-Induced CD8 T Cells in Adipose Tissue Mediate Macrophages to Sustain Low-Grade Chronic Inflammation. Front. Immunol. 2021 Jun 23;12:680944.
69. Kleer CG, Cao Q, Varambally S, Shen R, Ota I, Tomlins SA, et al. EZH2 is a marker of aggressive breast cancer and promotes neoplastic transformation of breast epithelial cells. Proc. Natl. Acad. Sci. U. S. A. 2003 Sep 30;100(20):11606–11.
70. Kooistra SM, Helin K. Molecular mechanisms and potential functions of histone demethylases. Nat. Rev. Mol. Cell Biol. 2012 Apr 4;13(5):297–311.
71. Laplante M, Sabatini DM. An emerging role of mTOR in lipid biosynthesis. Curr. Biol. 2009 Dec 1;19(22):R1046–52.
72. Lauby-Secretan B, Scoccianti C, Loomis D, Grosse Y, Bianchini F, Straif K, et al. Body Fatness and Cancer--Viewpoint of the IARC Working Group. N. Engl. J. Med. 2016 Aug 25;375(8):794–8.
73. Lee K, Kruper L, Dieli-Conwright CM, Mortimer JE. The Impact of Obesity on Breast Cancer Diagnosis and Treatment. Curr. Oncol. Rep. 2019 Mar 27;21(5):41.
74. Lee Y-S, Lim K-H, Guo X, Kawaguchi Y, Gao Y, Barrientos T, et al. The Cytoplasmic Deacetylase HDAC6 Is Required for Efficient Oncogenic Tumorigenesis. Cancer Res. 2008 Spe 14;68(18):7561–9.
75. Li K, Shi H, Zhang B, Ou X, Ma Q, Chen Y, et al. Myeloid-derived suppressor cells as immunosuppressive regulators and therapeutic targets in cancer. Signal Transduct Target Ther. 2021a Oct 7;6(1):362.
76. Li S, Yu J, Huber A, Kryczek I, Wang Z, Jiang L, et al. Metabolism drives macrophage heterogeneity in the tumor microenvironment. Cell Rep. 2022 Apr 5;39(1):110609.
77. Li X, Dai A, Tran R, Wang J. Identifying miRNA biomarkers for breast cancer and ovarian cancer: a text mining perspective. Breast Cancer Res. Treat. 2023 Aug;201(1):5–14.
78. Li X, Liang Y, Lian C, Peng F, Xiao Y, He Y, et al. CST6 protein and peptides inhibit breast cancer bone metastasis by suppressing CTSB activity

and osteoclastogenesis. Theranostics. 2021b Oct 11;11(20):9821–32.
79. Łukasiewicz S, Czeczelewski M, Forma A, Baj J, Sitarz R, Stanisławek A. Breast Cancer-Epidemiology, Risk Factors, Classification, Prognostic Markers, and Current Treatment Strategies-An Updated Review. Cancers 2021 Aug 25;13(17). https://doi.org/10.3390/cancers13174287
80. Lukey MJ, Wilson KF, Cerione RA. Therapeutic strategies impacting cancer cell glutamine metabolism. Future Med. Chem. 2013 Sep;5(14):1685–700.
81. Mahmood N, Arakelian A, Muller WJ, Szyf M, Rabbani SA. An enhanced chemopreventive effect of methyl donor S-adenosylmethionine in combination with 25-hydroxyvitamin D in blocking mammary tumor growth and metastasis. Bone Res. 2020 Jul 22;8:28.
82. Mahmood N, Cheishvili D, Arakelian A, Tanvir I, Khan HA, Pépin A-S, et al. Methyl donor S-adenosylmethionine (SAM) supplementation attenuates breast cancer growth, invasion, and metastasis in vivo; therapeutic and chemopreventive applications. Oncotarget. 2018 Jan 12;9(4):5169–83.
83. Mauro C, Smith J, Cucchi D, Coe D, Fu H, Bonacina F, et al. Obesity-Induced Metabolic Stress Leads to Biased Effector Memory CD4 T Cell Differentiation via PI3K p110δ-Akt-Mediated Signals. Cell Metab. 2017 Mar 7;25(3):593–609.
84. Misra S, Hascall VC, Markwald RR, Ghatak S. Interactions between Hyaluronan and Its Receptors (CD44, RHAMM) Regulate the Activities of Inflammation and Cancer. Front. Immunol. 2015 May 6;6:201.
85. Moore LD, Le T, Fan G. DNA methylation and its basic function. Neuropsychopharmacology. 2013 Jan;38(1):23–38.
86. Navas LE, Carnero A. NAD metabolism, stemness, the immune response, and cancer. Signal Transduct Target Ther. 2021 Jan 1;6(1):2.
87. Na YK, Hong HS, Lee DH, Lee WK, Kim DS. Effect of body mass index on global DNA methylation in healthy Korean women. Mol. Cells. 2014 Jun;37(6):467–72.
88. Neuhouser ML, Chlebowski RT, Anderson GL. Association Between Obesity and Postmenopausal Breast Cancer Risk-Reply. JAMA Oncol. 2015. p. 1171.
89. Pauken KE, Sammons MA, Odorizzi PM, Manne S, Godec J, Khan O, et al. Epigenetic stability of exhausted T cells limits durability of reinvigoration by PD-1 blockade. Science. 2016 Dec2;354(6316):1160–5.
90. Petrus P, Bialesova L, Checa A, Kerr A, Naz S, Bäckdahl J, et al. Adipocyte Expression of SLC19A1 Links DNA Hypermethylation to Adipose Tissue Inflammation and Insulin Resistance. J. Clin. Endocrinol. Metab. 2018 Feb 1;103(2):710–21.
91. Picon-Ruiz M, Morata-Tarifa C, Valle-Goffin JJ, Friedman ER, Slingerland JM. Obesity and adverse breast cancer risk and outcome: Mechanistic insights and strategies for intervention. CA Cancer J. Clin. 2017 Sep;67(5):378–97.
92. Porsche CE, Delproposto JB, Geletka L, O'Rourke R, Lumeng CN. Obesity results in adipose tissue T cell exhaustion. JCI Insight 2021 Apr 22;6(8). https://doi.org/10.1172/jci.insight.139793
93. Purnell JQ. Definitions, Classification, and Epidemiology of Obesity. Endotext. MDText.com, Inc.; 2023.
94. Randle PJ, Garland PB, Newsholme EA, Hales CN. The glucose fatty acid cycle in obesity and maturity onset diabetes mellitus. Ann. N. Y. Acad. Sci. 1965 Oct 8;131(1):324–33.
95. Ray Chaudhuri A, Nussenzweig A. The multifaceted roles of PARP1 in DNA repair and chromatin remodeling. Nat. Rev. Mol. Cell Biol. 2017 Oct;18(10):610–21.
96. Rifaï K, Judes G, Idrissou M, Daures M, Bignon Y-J, Penault-Llorca F, et al. SIRT1-dependent epigenetic regulation of H3 and H4 histone acetylation in human breast cancer. Oncotarget. 2018 Jul 17;9(55):30661–78.
97. Rios Garcia M, Steinbauer B, Srivastava K, Singhal M, Mattijssen F, Maida A, et al. Acetyl-CoA Carboxylase 1-Dependent Protein Acetylation Controls Breast Cancer Metastasis and Recurrence. Cell Metab. 2017 Dec 5;26(6):842–55.e5.
98. Roll JD, Rivenbark AG, Jones WD, Coleman WB. DNMT3b overexpression contributes to a hypermethylator phenotype in human breast cancer cell lines. Mol. Cancer. 2008 Jan 25;7:15.
99. Rotili D, Mai A. Targeting Histone Demethylases: A New Avenue for the Fight against Cancer. Genes Cancer. 2011 Jun;2(6):663–79.
100. Safarzadeh E, Mohammadi A, Mansoori B, Duijf PHG, Hashemzadeh S, Khaze V, et al. STAT3 Silencing and TLR7/8 Pathway Activation Repolarize and Suppress Myeloid-Derived Suppressor Cells From Breast Cancer Patients. Front. Immunol. 2020;11:613215.
101. Samaržija I, Tomljanović M, Novak Kujundžić R, Trošelj KG. EZH2 Inhibition and Cisplatin as a Combination Anticancer Therapy: An Overview of Preclinical Studies. Cancers 2022 Sep 29;14(19). https://doi.org/10.3390/cancers14194761
102. Santonja A, Sánchez-Muñoz A, Lluch A, Chica-Parrado MR, Albanell J, Chacón JI, et al. Triple negative breast cancer subtypes and pathologic complete response rate to neoadjuvant chemotherapy. Oncotarget. 2018 May 29; 9(41):26406–16.
103. Sckisel GD, Bouchlaka MN, Monjazeb AM, Crittenden M, Curti BD, Wilkins DEC, et al. Out-of-Sequence Signal 3 Paralyzes Primary CD4(+) T-Cell-Dependent Immunity. Immunity. 2015 Aug 18;43(2):240–50.
104. Sen DR, Kaminski J, Barnitz RA, Kurachi M, Gerdemann U, Yates KB, et al. The epigenetic landscape of T cell exhaustion. Science. 2016 Dec 2;354(6316):1165–9.

105. Serefidou M, Venkatasubramani AV, Imhof A. The Impact of One Carbon Metabolism on Histone Methylation. Front. Genet. 2019 Aug 28;10:764.
106. Sharda A, Rashid M, Shah SG, Sharma AK, Singh SR, Gera P, et al. Elevated HDAC activity and altered histone phospho-acetylation confer acquired radio-resistant phenotype to breast cancer cells. Clin. Epigenetics. 2020 Jan 3;12(1):4.
107. Shi J, Wang Y, Zeng L, Wu Y, Deng J, Zhang Q, et al. Disrupting the interaction of BRD4 with diacetylated Twist suppresses tumorigenesis in basal-like breast cancer. Cancer Cell. 2014 Feb 10; 25(2):210–25.
108. Shi L, Tu BP. Acetyl-CoA and the regulation of metabolism: mechanisms and consequences. Curr. Opin. Cell Biol. 2015 Apr;33:125–31.
109. Shirakawa K, Sano M. Drastic transformation of visceral adipose tissue and peripheral CD4 T cells in obesity. Front. Immunol. 2022;13:1044737.
110. Sido JM, Yang X, Nagarkatti PS, Nagarkatti M. Δ9-Tetrahydrocannabinol-mediated epigenetic modifications elicit myeloid-derived suppressor cell activation via STAT3/S100A8. J. Leukoc. Biol. 2015 Apr;97(4):677–88.
111. Smith SG, Sestak I, Morris MA, Harvie M, Howell A, Forbes J, et al. The impact of body mass index on breast cancer incidence among women at increased risk: an observational study from the International Breast Intervention Studies. Breast Cancer Res. Treat. Springer; 2021 Mar 3;188(1):215–23.
112. Suganuma T, Workman JL. Chromatin and Metabolism. Annu. Rev. Biochem. 2018 Jun 20;87:27–49.
113. Takahashi H, McCaffery JM, Irizarry RA, Boeke JD. Nucleocytosolic acetyl-coenzyme a synthetase is required for histone acetylation and global transcription. Mol. Cell. 2006 Jul 21;23(2):207–17.
114. Tao M-H, Marian C, Nie J, Ambrosone C, Krishnan SS, Edge SB, et al. Body mass and DNA promoter methylation in breast tumors in the Western New York Exposures and Breast Cancer Study. Am. J. Clin. Nutr. 2011 Sep;94(3):831–8.
115. Terranova-Barberio M, Thomas S, Ali N, Pawlowska N, Park J, Krings G, et al. HDAC inhibition potentiates immunotherapy in triple negative breast cancer. Oncotarget. 2017 Dec 26;8(69):114156–72.
116. Teslow EA, Mitrea C, Bao B, Mohammad RM, Polin LA, Dyson G, et al. Obesity-induced MBD2_v2 expression promotes tumor-initiating triple-negative breast cancer stem cells. Mol. Oncol. 2019 Apr;13(4):894–908.
117. Thakur C, Qiu Y, Fu Y, Bi Z, Zhang W, Ji H, et al. Epigenetics and environment in breast cancer: New paradigms for anti-cancer therapies. Front. Oncol. 2022 Sep 15;12:971288.
118. Tietscher S, Wagner J, Anzeneder T, Langwieder C, Rees M, Sobottka B, et al. A comprehensive single-cell map of T cell exhaustion-associated immune environments in human breast cancer. Nat. Commun. 2023 Jan 6;14(1):98.
119. Tsang DPF, Cheng ASL. Epigenetic regulation of signaling pathways in cancer: role of the histone methyltransferase EZH2. J. Gastroenterol. Hepatol. 2011 Jan;26(1):19–27.
120. Turcot V, Tchernof A, Deshaies Y, Pérusse L, Bélisle A, Marceau S, et al. LINE-1 methylation in visceral adipose tissue of severely obese individuals is associated with metabolic syndrome status and related phenotypes. Clin. Epigenetics. 2012 Jul 2;4(1):10.
121. Veglia F, Sanseviero E, Gabrilovich DI. Myeloid-derived suppressor cells in the era of increasing myeloid cell diversity. Nat. Rev. Immunol. 2021 Aug;21(8):485–98.
122. van Veldhoven K, Polidoro S, Baglietto L, Severi G, Sacerdote C, Panico S, et al. Epigenome-wide association study reveals decreased average methylation levels years before breast cancer diagnosis. Clin. Epigenetics. 2015 Aug 4;7(1):67.
123. Wang C, Chen X, Liu X, Lu D, Li S, Qu L, et al. Discovery of precision targeting EZH2 degraders for triple-negative breast cancer. Eur. J. Med. Chem. 2022a Aug 5;238:114462.
124. Wang C, Qu L, Li S, Yin F, Ji L, Peng W, et al. Discovery of First-in-Class Dual PARP and EZH2 Inhibitors for Triple-Negative Breast Cancer with Wild-Type BRCA. J. Med. Chem; 2021 64, 12630
125. Wang J, Wang Q, Guan Y, Sun Y, Wang X, Lively K, et al. Breast cancer cell-derived microRNA-155 suppresses tumor progression via enhancing immune cell recruitment and antitumor function. J. Clin. Invest. 2022b Oct 3;132(19). https://doi.org/10.1172/JCI157248
126. Wang X, Cao Q, Yu L, Shi H, Xue B, Shi H. Epigenetic regulation of macrophage polarization and inflammation by DNA methylation in obesity. JCI Insight. 2016 Nov 17;1(19):e87748.
127. Wang X, Zhang Y, Wang S, Ni H, Zhao P, Chen G, et al. The role of CXCR3 and its ligands in cancer. Front. Oncol. 2022c Nov 21;21;12:1022688.
128. Wang Y, Sun Z, Szyf M. S-adenosyl-methionine (SAM) alters the transcriptome and methylome and specifically blocks growth and invasiveness of liver cancer cells. Oncotarget. 2017 Dec 19;8(67):111866–81.
129. Wang Z, Aguilar EG, Luna JI, Dunai C, Khuat LT, Le CT, et al. Paradoxical effects of obesity on T cell function during tumor progression and PD-1 checkpoint blockade. Nat. Med. 2019 Jan;25(1):141–51.
130. Wei Y, Xiang H, Zhang W. Review of various NAMPT inhibitors for the treatment of cancer. Front. Pharmacol. 2022 Sep 7;13:970553.
131. Wellen KE, Hatzivassiliou G, Sachdeva UM, Bui TV, Cross JR, Thompson CB. ATP-citrate lyase links cellular metabolism to histone acetylation. Science. 2009 May 22;324(5930):1076–80.
132. Widschwendter P, Friedl TWP, Schwentner L, DeGregorio N, Jaeger B, Schramm A, et al. The influence of obesity on survival in early, high-risk breast cancer: results from the randomized SUCCESS A trial. Breast Cancer Res. BioMed Central; 2015 Sep 18;17(1):1–11.

133. Wong EM, Southey MC, Terry MB. Integrating DNA methylation measures to improve clinical risk assessment: are we there yet? The case of BRCA1 methylation marks to improve clinical risk assessment of breast cancer. Br. J. Cancer. Nature Publishing Group; 2020 Feb 18;122(8):1133–40.
134. Wu X, Li J, Chang K, Yang F, Jia Z, Sun C, et al. Histone H3 methyltransferase Ezh2 promotes white adipocytes but inhibits brown and beige adipocyte differentiation in mice. Biochim. Biophys. Acta Mol. Cell Biol. Lipids. 2021 Jun;1866(6):158901.
135. Yamaguchi H, Du Y, Nakai K, Ding M, Chang S-S, Hsu JL, et al. EZH2 contributes to the response to PARP inhibitors through its PARP-mediated poly-ADP ribosylation in breast cancer. Oncogene. 2018 Jan 11;37(2):208–17.
136. Yang Q, Zhao S, Shi Z, Cao L, Liu J, Pan T, et al. Chemotherapy-elicited exosomal miR-378a-3p and miR-378d promote breast cancer stemness and chemoresistance via the activation of EZH2/STAT3 signaling. J. Exp. Clin. Cancer Res. 2021 Apr 6;40(1):120.
137. Yunna C, Mengru H, Lei W, Weidong C. Macrophage M1/M2 polarization. Eur. J. Pharmacol. 2020 Jun 15;877:173090.
138. Zatterale F, Raciti GA, Prevenzano I, Leone A, Campitelli M, De Rosa V, et al. Epigenetic Reprogramming of the Inflammatory Response in Obesity and Type 2 Diabetes. Biomolecules 2022 Jul 14;12(7). https://doi.org/10.3390/biom12070982
139. Zhang H, Zhang N, Liu Y, Su P, Liang Y, Li Y, et al. Epigenetic Regulation of by Drives Metastatic Progression in Triple-Negative Breast Cancer. Cancer Res. 2019 Jul 1;79(13):3347–59.
140. Zhang S, Dang Y. Roles of gut microbiota and metabolites in overweight and obesity of children. Front. Endocrinol. . 2022 Sep 8;13:994930.
141. Zhao Y, Lin L, Li J, Xiao Z, Chen B, Wan L, et al. CD4 T cells in obesity and obesity-associated diseases. Cell. Immunol. 2018 Oct;332:1–6.
142. Zhuang J, Huo Q, Yang F, Xie N. Perspectives on the Role of Histone Modification in Breast Cancer Progression and the Advanced Technological Tools to Study Epigenetic Determinants of Metastasis. Front. Genet. 2020 Oct 29;11:603552.
143. Zhu Y, Guo M, Zhang L, Xu T, Wang L, Xu G. Biomarker triplet NAMPT/VEGF/HER2 as a de novo detection panel for the diagnosis and prognosis of human breast cancer. Oncol. Rep. 2016 Jan;35(1):454–62.
144. Rey, M., Irondelle, M., Waharte, F., Lizarraga, F., & Chavrier, P. (2011). HDAC6 is required for invadopodia activity and invasion by breast tumor cells. European journal of cell biology, Feb 1;90(2–3), 128–135.
145. Vasantharajan, S.S., Eccles, M.R., Rodger, E.J., Pattison, S., McCall, J.L., Gray, E.S., Calapre, L. and Chatterjee, A., 2021. The epigenetic landscape of circulating tumour cells. Biochimica et Biophysica Acta (BBA)-Reviews on Cancer, Apr 1;1875(2), p.188514.

# Epigenetic Modulations by Microbiome in Breast Cancer

4

Yuanji Zhao and Sanchita Bhatnagar

**Abstract**

Recent studies have identified a critical role of the diverse and dynamic microbiome in modulating various aspects of host physiology and intrinsic processes. However, the altered microbiome has also become a hallmark of cancer, which could influence the tumor microenvironment. Aberrations in epigenetic regulation of tumor suppressors, apoptotic genes, and oncogenes can accentuate breast cancer onset and progression. Interestingly, recent studies have established that the microbiota modulates the epigenetic mechanisms at global and gene-specific levels. While the mechanistic basis is unclear, the cross-talk between the microbiome and epigenetics influences breast cancer trajectory. Here, we review different epigenetic mechanisms of mammalian gene expression and summarize the host-associated microbiota distributed across the human body and their influence on cancer and other disease-related genes. Understanding this complex relationship between epigenetics and the microbiome holds promise for new insights into effective therapeutic strategies for breast cancer patients.

**Keywords**

Microbiota · Microbial diversity · Metagenomic, metabolomic, lipidomic, and metatranscriptomic · Dietary patterns · Normobiosis · RNA sequencing

## 4.1 Introduction to Epigenetic Regulation of Gene Expression

Differential gene expression is essential for cellular differentiation, growth, development, and sustained life. In addition to genetic alterations, the epigenetic modifications of DNA regulate mammalian gene expression. Epigenetic modifications are reversible and can be inherited through generations. The epigenetic mechanisms modify DNA without changing the sequence through DNA methylation, chromatin remodeling, histone modifications, and small noncoding RNAs (ncRNA). DNA methylation is a heritable epigenetic mark involving the covalent transfer of a methyl group to the C-5 position of the cytosine ring of DNA in CpG islands catalyzed by DNA methyltransferases (DNMTs). The methylation of DNA at the promoter, insulator, or enhancer defines different stages of human devel-

Y. Zhao · S. Bhatnagar (✉)
Department of Medical Microbiology and Immunology, University of California Davis School of Medicine, Davis, CA, USA
e-mail: sbbhatnagar@ucdavis.edu



S. Bhatnagar (ed.), *Epigenetic Mechanisms in Breast Cancer Therapy and Resistance*, Advances in Experimental Medicine and Biology 1465, https://doi.org/10.1007/978-3-031-66686-5_4

opment [1], silences repetitive elements [2], prevents the integration of viral sequences [3], facilitates genomic imprinting [4], inactivates X chromosome in females [5], and regulates transcription [6].

Strikingly, numerous studies identified the associations between altered DNA methylation states of tumor suppressors and oncogenes with tumorigenesis [7] and metastasis [8]. For example, hypermethylation of Breast Cancer Type 1 Susceptibility Protein (BRCA1) is an early event in triple-negative breast cancer (TNBC), occurring at a higher frequency than the BRCA1 pathogenic variant [9]. Interestingly, the BRCA1 promoter hypermethylation correlates with a homologous recombination deficiency, immune cell-type signatures, genome-wide DNA methylation, and transcriptional phenotype similar to TNBC tumors with *BRCA1*-inactivating variants. Another early tumorigenic marker is hypermethylation of Phosphatase and Tensin Homolog (PTEN), reported in ~34% of breast cancers [10]. PTEN methylation is significantly associated with poor progression-free survival (PFS) and worse overall survival (OS) in breast cancer patients [10].

Nucleosomes are the core unit of DNA that includes DNA wrapped around histone proteins, often subjected to covalent modifications affecting the physical state of chromatin and the corresponding transcriptional states. Posttranslational modifications (PTMs) of histones are essential regulatory mechanisms that modulate gene expression through covalent changes in histone proteins, such as methylation, phosphorylation, acetylation, ubiquitylation, SUMOylation, glycosylation, and ADP-ribosylation [11]. Methylation of specific histone residues can either activate or repress gene expression [12], while phosphorylation can regulate chromatin condensation and gene activation [13]. Acetylation neutralizes the positive charge of histones, relaxing the chromatin structure and promoting the gene transcription [14]. Ubiquitylation and SUMOylation control protein stability, nuclear localization, and protein-protein interactions [15]. Glycosylation adds sugar moieties to histones, influencing chromatin organization and cellular processes [16]. ADP-ribosylation affects DNA repair, transcription, and chromatin remodeling [17].

While PTMs orchestrate the intricate chromatin dynamics underlying gene regulation and cellular function, these modifications are catalyzed by a family of histone-modifying enzymes. Histone-modifying enzymes, referred to as the writers, catalyze the addition of the epigenetic marks. Specifically, the histone methyltransferases (HMTs) add the methyl mark, and the histone acetyltransferases (HATs) catalyze the acetylation of histones. The histone lysine methyltransferases (HKMTs) and histone protein arginine methyltransferases (PRMTs) catalyze the transfer of methyl groups S-adenosyl methionine (SAM) donor to the Ɛ-amino or ω-guanidino groups of lysine or arginine residues respectively. On the other hand, the HATs use acetyl-CoA as a cofactor to acetylate an Ɛ-amino side chain of lysine residues in histones. The HMTs and HATs are associated with mutations, genetic translocations, and altered gene expression in cancer, developmental disorders, and other pathologies. For example, the histone 3 at lysine 36 (H3K36)-specific methyltransferases NSD2 directly interacts with NF-kB to activate target genes, such as IL-6, IL-8, vascular endothelial growth factor A (VEGFA), cyclin D, Bcl-2, and surviving in prostate cancer cells [18]. The NF-kB and NSD2 partnering elevates histone H3K36me2 and H3K36me3 marks in prostate tumors, directly implicating NSD2 methyltransferase activity in the gene deregulation [19].

Chromatin remodeling complexes modulate gene expression by moving, sliding, disrupting, or restructuring the nucleosome. All chromatin remodelers contain a highly conserved ATPase domain that hydrolyzes ATP to disrupt histone-DNA contacts. The four families of chromatin remodeling complexes include the switching defective/sucrose nonfermenting (SWI/SNF), the imitation switch (ISWI), the chromodomain helicase DNA binding (CHD), and the inositol requiring 80 (INO80). Chromatin remodeling involves dissociating genomic DNA at the edge of the nucleosome, modulation, and repositioning of DNA on the histone octamer without changing the total number of histone-DNA contacts.

ncRNAs have diverse functions in regulating gene expression at the translational, transcriptional, or posttranscriptional levels. ncRNAs regulate transcription by recruiting chromatin-remodeling and repressive complexes, such as polycomb complex proteins (PcG). PcG is a multi-protein complex, polycomb repressive complex 1 (PRC1) and complex 2 (PRC2), at the center of chromatin-based transcriptional repression. PRC1 compacts chromatin and monoubiquitinates histone H2A at lysine 119 residues. PRC2 also compacts chromatin and catalyzes the methylation of histone H3 at lysine 27. One of the most studied long ncRNAs is X inactive specific transcript *(Xist),* a key regulator of X chromosome inactivation in eutherian mammals [20]. *Xist* evades export into the cytoplasm and instead coats the X chromosome in *cis* to accumulate repressive histone marks, such as trimethylation of H3K27 (H3K27me3), monoubiquitination of Histone 2A (H2Aub), and methylation of DNA. The "coating" of the silenced chromatin region provided the first model of how long ncRNAs might contribute to a stable epigenetic gene silencing [21].

## 4.2 Influence of Environmental Exposures on Epigenetic Mechanisms

Environmental exposures, such as diet, stress, behavior, physical activity, and working habits, affect host-cell physiology and intrinsic processes by altering epigenetic mechanisms (summarized in Fig. 4.1). Also, prenatal and postnatal environmental exposures influence global and gene-specific epigenetic changes. Maternal exposure to drugs, stress, and toxicants reprograms epigenetic regulation, which increases the likelihood of cancer and neurodevelopmental and behavioral deficits in the offspring [22, 23]. Arsenic exposure causes DNA methylation-dependent transcriptional silencing of tumor suppressors, such as *p53*, *CDKN2A* *($p16^{INK4A}$)*, Ras association domain family member 1 (*RASSF1A*), and death-associated protein kinase *(DAPK)* [24–26]. Notably, arsenate also induces carcinogenesis by altering histone methylations representing gene silencing and activating marks, such as H3K9me2, H3K27me3, and H3K4me3 [27]. Arsenate also induces microRNA expression, such as miR190, to regulate the gene expression [28]. In mice, inhalational exposure to 1,3-butadiene causes gene-specific methylation changes in histones H3K9, H3K27, and H4K20 [29]. Perinatal exposure to diethylstilbestrol causes DNA methylation-dependent transcriptional activation of several critical cancer-related genes, including lactoferrin (Lf), nucleosomal binding protein 1 (Nsbp1), and c-fos [30–32]. Most of the alteration of DNA methylation is the direct effect of exposures on the function of DNA methyltransferases, Dnmt1, Dnmt3a, and Dnmt3b [33].

Likewise, DNA methylation is also sensitive to stress in early development and adulthood. Maternal separation anxiety is linked to long-lasting stress on adult health through dysregulation of epigenetic processes, such as altered DNA methylation levels across promoters of Nuclear Receptor *Subfamily 3 group C member 1 (Nr3c1), Arginine* Vasopressin *(Avp) and* Nuclear Receptor *Subfamily 4 group A member 1 (Nr4a1)* in the hippocampus [34–36]. The social isolation of adult mice led to distinct epigenetic patterns in *HDAC1*, *HDAC3*, and SLC6A4, where gene-specific effects on DNA methylation were identified [37]. Experimental evidence also revealed aberrant DNA methylation in genes regulating the hypothalamic-pituitary-adrenal stress axis, such as Nr3c1, Avp, Crh, and Crh2 [38–41], and neurotrophic genes, such as BDNF, 5HT1A, and Gad1 [42–44]. The changes in DNA methylation are often dynamic and accompanied by changes at multiple gene loci. For example, the fear conditioning transcriptionally alters the memory suppressor gene PP1 through rapid methylation but demethylation of Reelin [45]. The hypermethylation of the glucocorticoid receptor gene and its reduced expression were identified in suicide victims and associated with childhood abuse [46].

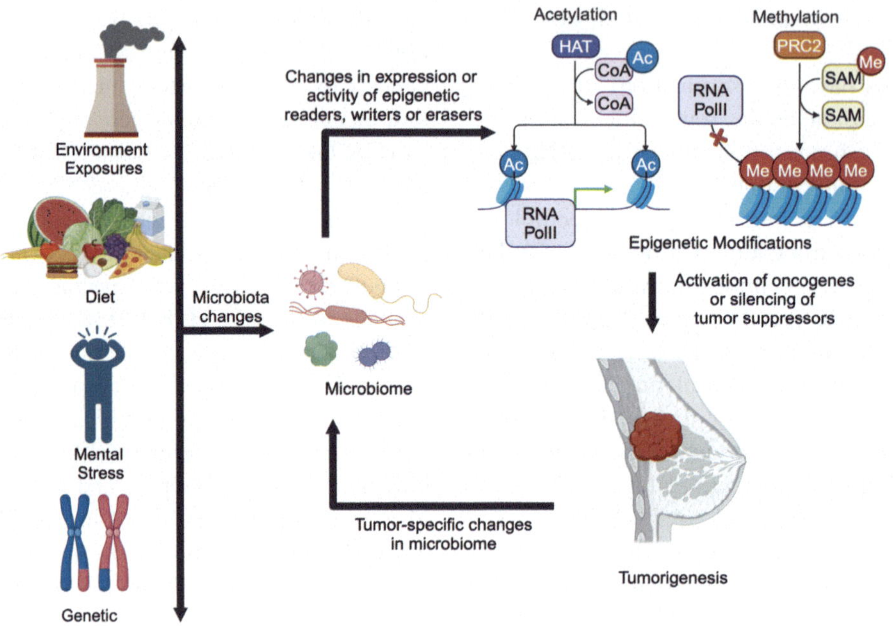


**Fig. 4.1** Influence of environmental exposures on epigenetic mechanisms through the microbiome. Schematic showing how different environmental exposures, diet, stress, and genetic factors mediate their effects on epigenetic readers, writers, and erasers. Aberrant histone modifications are linked to tumorigenesis and progression

## 4.3 Microbiome Diversity Across Human Body

The human microbiome is diverse and constitutes 10–100 trillion symbiotic microbes distributed in nearly every human body site [47]. While the potential application of microbiome in personalized medicine is a relatively new and emerging field, Antonie van Leeuwenhoek was the first to study the diversity of oral and fecal microbiota in the 1680s [48]. The NIH Human Microbiome Project (HMP) mapped and characterized the human microbiome in terms of its function in human health and disease [49]. Notably, there are 48 main microbial habitats in the human body, with 34% microbial habitats in human skin, 25% of all habitats in the gastrointestinal tract, and 20% in the cavities of the head and neck region [49]. Most human microbiome consists of bacteria, fungi, viruses, archaea, and protozoa, especially in the gut and oral cavity (summarized in Fig. 4.2).

The bacterial biofilms feature of the oral cavity has provided significant insight into the bacterial communities and their ability to modulate the pH, redox, and oxygen levels. The most predominant bacterial species inhabiting the oral cavity include *Gemella*, *Granulicatella*, *Streptococcus*, *Veillonella genera*, *Actinomyces* spp., *Streptococcus* spp., *F. nucleatum*, *Tannerella* spp., *Rothia* spp., and *Neisseria* spp. [50]. The majority of viruses in the oral cavity of healthy individuals mainly include bacteriophages, such as Siphoviridae, Myoviridae, and Podoviridae [51–53]. Additionally, the healthy human oral cavity also hosts fungi and yeast species [50], which include *Candida albicans*, *C. glabrata*, *C. parapsilosis*, *C. tropicalis*, *C. krusei*, *C. stellatoidea*, *C. kefyr*, *C. khmerensis*, and *C. metapsilosis* [54].

The esophageal mucosa also exhibits a high degree of compositional overlap with the oral cavity, including *Streptococcus*, *Prevotella*, and *Veillonella* [55]. While the esophagus is exposed to many microbes through ingested food in transit to the stomach, the microbes are largely transient. The benefits of the oral microbiome are multifold, which include aiding in food digestion, nutrient absorption, the maturity of the

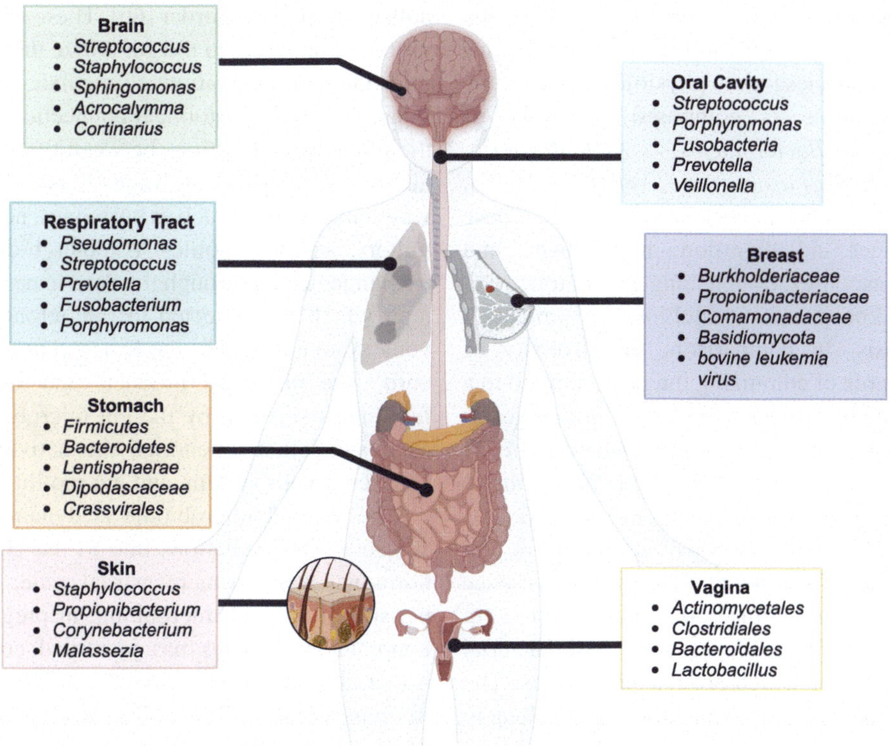


**Fig. 4.2** Microbiome diversity across different human tissues

host's mucosa and immune system, metabolism and fat storage, processing, and eliminating environmental toxins. The oral microbiome also balances pro-inflammatory and anti-inflammatory processes, preventing the colonization and growth of disease-causing microorganisms and promoting overall health and well-being.

Once considered sterile, the lung was recently identified as a host site for a dynamic and diverse microbiota owing to new methods and technologies. The lung microbiota can be further stratified by the anatomy, where the upper respiratory tract consists of the nasal cavity, paranasal sinuses, pharynx, and supraglottic portion of the larynx and lower respiratory tract, which comprises the trachea and lungs. The dominant taxa in the upper respiratory tract include *Moraxella*, *Staphylococcus*, *Corynebacterium*, *Haemophilus*, and *Streptococcus species*. In contrast, the oropharynx demonstrates a high abundance of *Prevotella*, *Veillonella*, *Streptococcus*, *Leptotrichia*, *Rothia*, *Neisseria*, and *Haemophilus species*. The lower respiratory tract has a relatively low microbial biomass due to the rapid microbial clearance by physiological mechanisms.

A recent 16S ribosomal RNA sequencing of the breast tissue from women revealed abundant and diverse taxa that include *Bacillus, Acinetobacter, Enterobacteriaceae, Pseudomonas, Staphylococcus, Propionibacterium, Comamonadaceae, Gammaproteobacteria, Listeria welshimeri, and Prevotella* [56]. Fungi in healthy breast tissue include *Basidiomycota* and *Ascomycota* [57]. Various viruses, such as bovine leukemia virus (BLV), mouse mammary tumor virus, high-risk human papillomavirus, and Epstein-Barr virus [58], can also be found.

Our understanding of the human gut microbiome and its role in diseases has expanded since the development of molecular tools and techniques, such as metagenomic, metabolomic, lipidomic, and metatranscriptomic. Dietary patterns and environmental factors shape the dynamic and diverse nature of gut microbiota. While the microbiota composition and relative populations of bacterial species link to physiologic health, the gut microbiome is composed primarily of *Firmicutes*, *Bacteroidetes*, *Actinobacteria*, *Proteobacteria*, *Fusobacteria*, *Verrucomicrobia*, *Tenericutes*, and *Lentisphaerae* [58]. These microbiomes aid digestion, metabolism, and immune responses, establishing a symbiotic relationship known as normobiosis to maintain homeostasis. In comparison, the diversity of fungi capable of colonizing the gut is limited to a small number of species, such as *Candida* yeasts and yeasts in the family *Dipodascaceae* (*Galactomyces*, *Geotrichum*, and *Saprochaete*) [59]. *Malassezia* and the filamentous fungus *Cladosporium* are also potential colonizers that may originate from the diet [59]. Viral presence can also be found in the human gut, with the gut virome consisting of eukaryotic and bacterial viruses, most of which are bacteriophages. The most abundant phages in the human gut belong to the *Crassvirales* order and infect bacterial species of the *Bacteroides genus*, such as *B. intestinalis* and *B. xylanisolvens* [60].

The bidirectional communication between the brain and gut is central to the brain-gut axis and occurs through the neural network, neuroendocrine system, immune system, and metabolic pathways. Neuroactive substances, such as serotonin, gamma-aminobutyric acid (GABA), SCFAs, and neurotoxins like ammonia and D-lactic acid, impact brain function [61]. Interestingly, changes in the gut microbiota can influence brain physiology and cognitive functions by controlling the synthesis and release of neurotransmitters. For example, *Bifidobacterium infantis* affects serotonin by raising tryptophan, which is converted to 5-hydroxytryptophan by tryptophan hydroxylase and converted to serotonin by aromatic amino acid decarboxylase [62]. *Lactobacillus reuteri* upregulates oxytocin in the plasma and brain [63]. *Lactobacillus rhamnosus* generates GABA and is implicated in depression and anxiety-like behavior in the murine model [64]. *Bifidobacterium longum* regulates vagus nerve signals [65], and *Bacteroides fragilis* regulates 4-ethylphenyl sulfate in a murine model of autism spectrum disorder [66]. These neurotransmitters originating in the gut utilize the brain and intestine communications to mediate their functions. The neuroanatomical connections mediate the intercommunications between the autonomic nervous system and the vagus nerve. This pathway senses stimuli such as hormones, neurotransmitters, and metabolites. Another bidirectional exchange occurs through the enteric nervous system, which is connected to the autonomic nervous system and the vagus nerve in the spinal cord. The microbial products and short-chain fatty acids secreted by the gut microbiota influence sympathetic neurons, the activity of the enteric nervous system, and gut motility.

The vaginal microbiome is dynamic that is subjected to fluctuations due to the changes in hormone levels during menstrual cycle. The comparison of vaginal microbiome in pregnant and nonpregnant females has identified contrasting bacterial populations. The *Lactobacillus* spp., *Actinomycetales*, *Clostridiales*, and *Bacteroidales* dominate the vagina of pregnant women [67]. In contrast, the predominance of *Lactobacillus* spp., *Actinobacteria*, *Prevotella*, *Veillonellaceae*, *Streptococcus*, *Proteobacteria*, *Bifidobacteriaceae*, *Bacteroides*, and *Burkholderiales* is observed in nonpregnant women [68]. Clinically, the abundance of *Lactobacilli* and acidic pH were considered signatures of a "healthy" vaginal microbiome. In contrast, a more diverse microbiome with higher pH has been correlated with preterm birth, "dysbiotic" conditions like bacterial vaginosis (BV), and the presence of vaginal pathogens like *Gardnerella* or *Trichomonas* [68].

## 4.4 Role of Microbiome in Breast Cancer

Breast cancer is the most common cancer type affecting women, with 287,850 diagnoses in the year 2022. It is classified based on the presence or absence of cell surface receptors like estrogen receptors (ER), progesterone receptors (PR), and human epidermal growth factor receptor 2 (HER2)—the lack of these three receptors results in triple-negative breast cancer. Breast cancer is a complex condition influenced by various factors such as age, lifestyle, exposure to carcinogens, and genetic predisposition [69]. Interestingly, 70% of breast cancer cases are diagnosed in individuals without any known risk factors [70]. Although early detection rates are improving and mortality rates have significantly declined, there is still a lack of consistency in predicting outcomes.

Recently, the gut microbiota has emerged as a new hallmark of cancer that influences breast cancer risk, response to treatment, and recurrence through metabolic pathways, cellular signaling, and immune response [71]. The therapeutically beneficial effects of microbiota supplementation have been revealed in clinics for different cancer types. The oral supplementation of *Akkermansia muciniphila* restored the efficacy of PD-1 blockade in an interleukin-12-dependent manner by increasing the recruitment of CCR9 + CXCR3 + CD4+ T lymphocytes into mouse tumor beds [72]. Treatment of melanoma patients with antibodies against CTLA-4 favored the outgrowth of *Bacillus fragilis* with anticancer properties [73]. The *Ruminococcaceae* family produces SCFAs that lead to the accumulation of $CD4^+$ T cells, inducible T-cell co-stimulator $(ICOS)^+$ T cells, and $CD8^+$ T cells to mediate their antitumor effects [74]. Furthermore, the gut microbiota is also linked to the increased regulatory T cell ($T_{reg}$) production or reduced differentiation of pathogenic T cells, thus preventing inflammatory diseases [74]. The $T_{regs}$ are characterized by $FOXP3^+/CD4^+$ expression, predominantly contribute to immune tolerance, and are associated with poorer prognosis [75]. $T_{reg}$ cell counts increase through different stages of breast cancer, from normal tissue to ductal carcinoma in situ (DCIS) and from DCIS to invasive carcinoma [76]. Notably, $T_{reg}$ cells expressing the FOXP3 transcription factor also regulate the immune response to commensal microbiota, and the metabolites produced by these bacteria could potentially regulate $T_{reg}$ cell turnover [77].

However, the shift in microbiome composition and expansion of otherwise rare species could also indicate a diseased state associated with tumor initiation and progression. Microbiomes can contribute to cancer development and progression through several mechanisms, including perturbation of the balance between cell proliferation and death, modulation of the host immune system, and impact on host metabolism. For example, the secretion of *B. fragilis* toxin (BFT) by *Bacteroides fragilis* in cancerous breast and gut is linked to epithelial hyperplasia in the mammary gland [78]. The altered microbiome can accelerate tumorigenesis by misregulating cellular proliferation rate and cell polarity through the Wnt/β-catenin pathway [79]. The bacterial secreted proteins such as FadA (Fusobacterium nucleatum), Avra (Salmonella typhi), and CagA (Helicobacter pylori) are linked to colorectal, hepatobiliary, and stomach cancer, respectively [80–82]. Colibactin from *Enterococaceae*, Bacteroides fragilis toxin (BFT) from *B. fragilis*, and cytolethal expansive toxin (CDT) from ε/γ *Proteobacteria* induce DNA damage that is characteristic of cancer cells [83].

With advancements in sequencing technologies and comprehensive, collaborative efforts, the microbiome has emerged as a potential determinant of breast cancer severity and activity. The composition of gut microbiota differs between breast cancer patients and various subtypes of breast cancer. For instance, the gut microbiota composition in patients with breast carcinoma differs from that in patients with benign tumors [84]. The malignant tumor group shows a significantly higher abundance of *Citrobacter* than the benign tumor group [85]. Pathway analysis of these groups revealed increased glycan and lipopolysaccharide biosynthesis in the malignant tumor group [86].

Furthermore, breast cancer patients exhibit different gut microbiota compositions in a stage-specific manner compared to healthy individuals (summarized in Fig. 4.3). The alpha diversity Shannon index was lower in fecal samples collected from early-stage breast cancer patients before treatment than in healthy individuals [87]. In particular, *Firmicutes*, *Verrucomicrobiaceae*, *Odoribacter* sp., and *Butyricimonas* sp. were enriched, but *Bacteroidetes* were in lower abundance [88]. For breast cancer patients undergoing

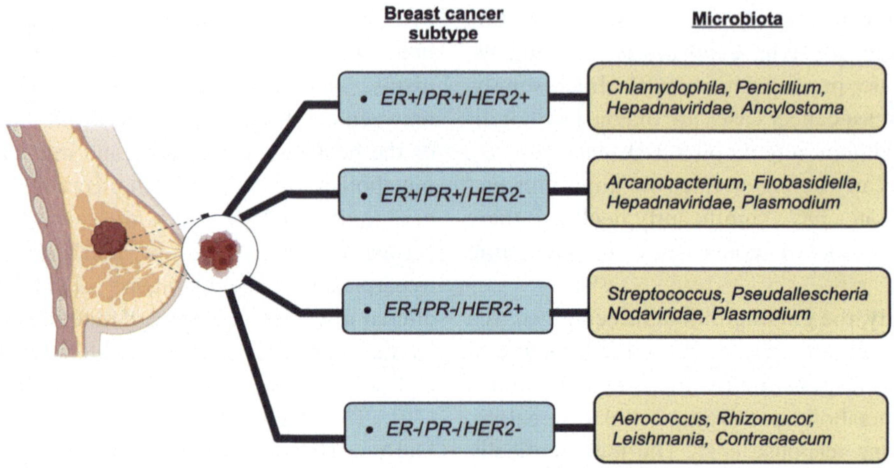


**Fig. 4.3** Association between microbiota composition and breast cancer subtypes

neoadjuvant chemotherapy, their fecal metabolic profiles are altered due to changes in the microbiota population. Bacterial metabolites like lactate and fumaric acid are downregulated in patients' post-chemotherapy [89], while SCFAs such as propionate, butyrate, and acetate increase in patients after three cycles of chemotherapy [90]. Another study on the intestinal microbiota in early breast cancer treatment demonstrated an association between the microbiome and prognosis in multiple types of early breast cancer [91]. Her2-positive and triple-positive breast cancer subtypes showed fewer bacterial (*Citrobacter*, *Streptobacillus*), fungal (*Enterocytozoon*, *Issatchenkia*), and parasitic (*Blastocystis*, *Cryptosporidium*, *Dientamoeba*, *Hartmannella*, *Isospora*, *Macracanthorhynchus*, *Metagonimus*) signatures [91]. Lower abundance of *Pseudoterranova*, *Ancylostoma*, *Trichuris*, and *Issatchenkia* correlated with increased disease-free time after treatment [91]. Aggressive tumors are characterized by a higher abundance of *Streptococcus* and *Lachnospiraceae* [92], while less aggressive tumors exhibit a higher abundance of specific health-related commensals [92]. Notably, specific bacterial species have been identified that may be associated with prognosis or overall health status in breast cancer patients. These findings indicate that fecal metabolites can change in correlation with the treatment period, and the gut microbiome can alter metabolic profiles to affect the risk of breast cancer.

Pathogenic breast tissue exhibits a microbiome that varies significantly from healthy tissue. *Proteobacteria*, *Actinobacteria*, and *Firmicutes* are the significant species identified in breast tissues. Proteobacteria is more prevalent in breast tumors, while Actinobacteria is more prevalent in healthy tissue [93]. When comparing microbiome composition within the same individual, breast tumors show lower bacterial richness than adjacent healthy tissue. *Actinobacteria* and *Propionibacteriaceae* are less abundant in tumor tissue, while *Proteobacteria*, *Firmicutes*, and *Alphaproteobacteria* are more abundant [87]. Moreover, a correlation between bacterial genera and malignancy has been observed when comparing breast tissue from women with malignant breast cancer to those with benign breast disease. A higher abundance of *Fusobacterium*, *Atopobium*, *Gluconacetobacter*, and others has been associated with malignancy [94]. Breast tissue microbiome also changes at different stages of breast cancer. As cancer stages progress, microbiota alterations can downregulate flavonoid biosynthesis and upregulate processes related to glycerophospholipids and ribosomes. Breast cancer patients also exhibit a higher relative abundance of *Bacillus*, *Enterobacteriaceae*, and *Staphylococcus* compared to normal breast tissue in healthy individuals [93]. These three genera were identified in normal breast tissue in previous research, suggesting that their higher concentration, rather than their mere presence, is associated with breast cancer incidence. Notably, E. coli and *Staphylococcus epidermidis*, known to produce the genotoxin colibactin that causes DNA double-strand breaks, have been hypothesized to promote breast cancer development [95].

The gut microbiome can be influenced by various factors, with diet being a significant component. The Mediterranean diet, for example, has been extensively studied in terms of microbiota and breast cancer. This diet, characterized by increased intake of monounsaturated and polyunsaturated fatty acids, gut microbiota-accessible carbohydrates, fruits, vegetables, and legumes, has been associated with improved health outcomes and a potential protective effect against breast cancer. Compared to a Western diet, the Mediterranean diet significantly affects the microbiota populations and metabolite profiles of mammary glands. The Mediterranean diet is associated with a tenfold higher abundance of *Lactobacillus* in breast tissue than mammary tissue from monkeys fed a Western diet, which is rich in saturated fats and sucrose and low in fiber [96]. Additionally, mammary tissue analysis from women primarily on the Mediterranean diet exhibited higher levels of bile acid metabolites and increased production of bacterial-processed bioactive compounds. Given that the increased bile acid metabolites are associated with increased apoptosis and repressed aromatase expression [97], it is speculated that metabolite-mediated activation of the farnesoid X receptor (NR1H4) signaling could favor anticancer properties.

## 4.5 The Microbiome-Directed Epigenetic Influence in Breast Cancer

Most recently, microbiota has been implicated in encoding environmental cues gestationally and transmitting them to the host through epigenetic changes. The link between the microbiome and epigenetic regulation is mainly through microbial metabolites (summarized in Fig. 4.4). The intestinal microbiota is dominated by *Firmicutes, Bacteroidetes, Actinobacteria*, and *Proteobacteria*, which produce a number of low-molecular-weight bioactive substances, such as folate, butyrate, biotin, and acetate. Several of these metabolites influence the availability of chemical donors for DNA or histone modifications [98]. Folate and B vitamins donate methyl groups for DNA or histone methylation. The microbial population also regulates the epigenetic modifying enzyme through expression or activity. Butyrate is a short-chain fatty acid that inhibits HDACs [99]. Butyrate produced by the cluster of *Firmicute* bacteria, comprised of *Clostridium leptum*, *Eubacterium rectale*, and *Clostridium coccoides*, is implicated in the transcriptional de-repression of cancer-related genes, such as p21 and BAK [100]; repression of angiogenesis; and suppression of pro-angiogenic factors, such as EGF and HIF-1$\alpha$ [101]. Biotin is a substrate for an epigenetic modification that involves covalent binding of biotin to histone proteins for gene silencing [102]. The gut microbiota metabolism can regulate the $NAD^+$/NADH levels, which controls histone deacetylase activity [103]. Several host-cell intrinsic processes directing epigenetic pathways are also regulated by the microbiome. For example, microbial colonization can induce HDAC3-mediated histone deacetylation on pro-inflammatory cytokine IL-12$\beta$ to repress inflammation and maintain anti-inflammatory activities [104]. The gut microbiota also regulates the absorption of cofactors of epigenetic modifiers, such as zinc, iodine, selenium, and cobalt. The isothiocyanates derived from sulfur-containing glucosinolates modulate DNA methylation. Specifically, sulforaphane downregulates DNMT1 to demethylate the telomerase reverse transcriptase [105].

Notably, the bacterial metabolites also have anticancer activity and, thus, could reduce cancer risk. Nisin produced by gram-positive *L. lactis* influences the innate immune system and modulates the chemokines [106]. A *Bifidobacterium pseudolongum* metabolite called inosine increases the efficacy of immune checkpoint blockade and demonstrates cryoprotective effects in breast cancer-induced hypoxia. Interestingly, metabolites also exert anticancer effects by mod-

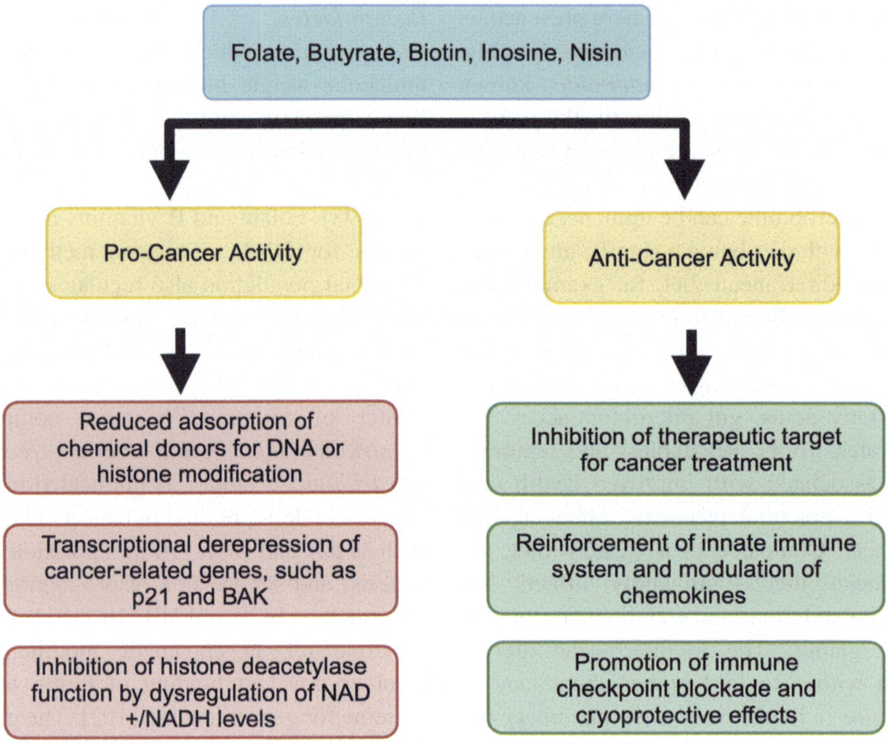


**Fig. 4.4** Bacterial metabolites exert anticancer and pro-cancer activity

ulating epigenetic mechanisms. The butyrate silences BAK and p21 by the activation of epigenetic mechanisms and by modulating the bioavailability of cofactors required by epigenetic modifiers [107]. HDACs are a therapeutic target owing to their function in metabolic pathways. As such, HDAC inhibitors are used in combinatorial therapies for cancer treatment. Butyrate and sodium butyrate are potent HDAC inhibitors demonstrating significant anticancer activity in breast cancer cell lines [108–110].

## 4.6 Approaches to Assess Microbiome

Given the significant impact of microbiome communities on breast cancer, it is crucial to assess the microbiome quantitatively and qualitatively. The main culture-independent techniques used for microbiome analysis include identifying the 16S ribosomal RNA and shotgun metagenomics. The 16S rRNA analysis is suitable for scaling up the analysis but provides limited taxonomical and functional resolution since it only captures part of the genomic content of a sample. In contrast, shotgun metagenomics allows for higher resolution by profiling microbial communities' taxonomic composition and functional potential, including the discovery of new bacterial genes and genomes [111, 112]. However, this approach is more expensive and requires more complex bioinformatic analyses. Notably, the approaches above only indicate the presence of microorganisms or genes and do not distinguish active from inactive members of a given microbiome. RNA sequencing (RNA-Seq) provides insights into active genes and pathways within a microbiome, but its experimental design is highly complex and has limitations. Moreover, it may only capture part of the metatranscriptome due to the high diversity and relative ratios of microbial communities as well as the short half-life of the RNA [113]. Other direct methodologies, such as scanning electron microscopy, can visualize microbial organization in fixed samples, but there are different methods of choice when identifying individual taxa in complex communities. Fluorescence in situ hybridization (FISH) using rRNA-targeted oligonucleotide probes can locate, visualize, and quantify microbial community members in fixed samples. However, it is limited in its ability to detect a small number of phylogenetically distinct target organisms simultaneously. Metabolomic and metaproteomic technologies utilizing mass spectrometry (MS) and nuclear magnetic resonance (NMR) spectrometry are also valuable approaches for characterizing the molecular profiles of the human microbiome. These techniques can quantify proteins and metabolites the microbiome produces, such as vitamins, bile salts, fatty acids, and polyphenols [114].

## 4.7 Conclusions

Cancer is a complex and multistep disease that is influenced by genetic, epigenetic, and environmental mechanisms. Recently, the microbiome has been identified as a new cancer hallmark with varied functions in healthy and diseased states, including cancer. We recognize that the diverse and dynamic nature of the microbiomes is at a crossroads with epigenetic and environmental factors. The technical advancement in sequencing tools and bioinformatics has helped better understand microbiota composition in different human tissues, its influence on health, and its contribution to disease onset and progression. Here, we have summarized the microbiome composition of the human body and discussed the association of microbiome-induced epigenetic changes with breast cancer. For future investigations, efforts should be focused on gaining a comprehensive understanding of microbiome structure and functions. To carry out informative and comprehensive microbiome research studies, the metagenomic analyses of patients supplemented with metatranscriptomics, metaproteomics, and metabolomics could provide important insights into microbiome-driven carcinogenesis.

**Acknowledgments** All figures are created and adapted with BioRender.com using an institutional license from the University of California Davis.

## References

1. Hon, G.C., et al., *Epigenetic memory at embryonic enhancers identified in DNA methylation maps from adult mouse tissues.* Nat Genet, 2013. **45**(10): p. 1198–206.
2. Yoder, J.A., C.P. Walsh, and T.H. Bestor, *Cytosine methylation and the ecology of intragenomic parasites.* Trends Genet, 1997. **13**(8): p. 335–40.
3. Zhang, D., S. Guo, and S.J. Schrodi, *Mechanisms of DNA Methylation in Virus-Host Interaction in Hepatitis B Infection: Pathogenesis and Oncogenetic Properties.* Int J Mol Sci, 2021. **22**(18).
4. Li, E., C. Beard, and R. Jaenisch, *Role for DNA methylation in genomic imprinting.* Nature, 1993. **366**(6453): p. 362–5.
5. Sharp, A.J., et al., *DNA methylation profiles of human active and inactive X chromosomes.* Genome Res, 2011. **21**(10): p. 1592–600.
6. Mattei, A.L., N. Bailly, and A. Meissner, *DNA methylation: a historical perspective.* Trends Genet, 2022. **38**(7): p. 676–707.
7. Jones, P.A. and M.L. Gonzalgo, *Altered DNA methylation and genome instability: a new pathway to cancer?* Proc Natl Acad Sci U S A, 1997. **94**(6): p. 2103–5.
8. Zhang, S., et al., *DNA methylation profiling to determine the primary sites of metastatic cancers using formalin-fixed paraffin-embedded tissues.* Nat Commun, 2023. **14**(1): p. 5686.
9. Glodzik, D., et al., *Comprehensive molecular comparison of BRCA1 hypermethylated and BRCA1 mutated triple negative breast cancers.* Nat Commun, 2020. **11**(1): p. 3747.
10. Ramadan, A., et al., *Clinical impact of PTEN methylation status as a prognostic marker for breast cancer.* J Genet Eng Biotechnol, 2021. **19**(1): p. 66.
11. Millan-Zambrano, G., et al., *Histone posttranslational modifications - cause and consequence of genome function.* Nat Rev Genet, 2022. **23**(9): p. 563–580.
12. Zhang, Y. and D. Reinberg, *Transcription regulation by histone methylation: interplay between different covalent modifications of the core histone tails.* Genes Dev, 2001. **15**(18): p. 2343–60.
13. Li, J., et al., *Programmable human histone phosphorylation and gene activation using a CRISPR/Cas9-based chromatin kinase.* Nat Commun, 2021. **12**(1): p. 896.
14. Nitsch, S., L. Zorro Shahidian, and R. Schneider, *Histone acylations and chromatin dynamics: concepts, challenges, and links to metabolism.* EMBO Rep, 2021. **22**(7): p. e52774.

15. Gill, G., *SUMO and ubiquitin in the nucleus: different functions, similar mechanisms?* Genes Dev, 2004. **18**(17): p. 2046–59.
16. Indellicato, R. and M. Trinchera, *Epigenetic Regulation of Glycosylation in Cancer and Other Diseases.* Int J Mol Sci, 2021. **22**(6).
17. Messner, S. and M.O. Hottiger, *Histone ADP-ribosylation in DNA repair, replication and transcription.* Trends Cell Biol, 2011. **21**(9): p. 534–42.
18. Kuo, A.J., et al., *NSD2 links dimethylation of histone H3 at lysine 36 to oncogenic programming.* Mol Cell, 2011. **44**(4): p. 609–20.
19. Yang, P., et al., *Histone methyltransferase NSD2/MMSET mediates constitutive NF-kappaB signaling for cancer cell proliferation, survival, and tumor growth via a feed-forward loop.* Mol Cell Biol, 2012. **32**(15): p. 3121–31.
20. Loda, A. and E. Heard, *Xist RNA in action: Past, present, and future.* PLoS Genet, 2019. **15**(9): p. e1008333.
21. Ponting, C.P., P.L. Oliver, and W. Reik, *Evolution and functions of long noncoding RNAs.* Cell, 2009. **136**(4): p. 629–41.
22. Kundakovic, M. and I. Jaric, *The Epigenetic Link between Prenatal Adverse Environments and Neurodevelopmental Disorders.* Genes (Basel), 2017. **8**(3).
23. Knopik, V.S., et al., *Prenatal substance exposure and offspring development: Does DNA methylation play a role?* Neurotoxicol Teratol, 2019. **71**: p. 50–63.
24. Cui, X., et al., *Chronic oral exposure to inorganic arsenate interferes with methylation status of p16INK4a and RASSF1A and induces lung cancer in A/J mice.* Toxicol Sci, 2006. **91**(2): p. 372–81.
25. Mass, M.J. and L. Wang, *Arsenic alters cytosine methylation patterns of the promoter of the tumor suppressor gene p53 in human lung cells: a model for a mechanism of carcinogenesis.* Mutat Res, 1997. **386**(3): p. 263–77.
26. Huang, Y.C., et al., *Sodium arsenite-induced DAPK promoter hypermethylation and autophagy via ERK1/2 phosphorylation in human uroepithelial cells.* Chem Biol Interact, 2009. **181**(2): p. 254–62.
27. Zhou, X., et al., *Arsenite alters global histone H3 methylation.* Carcinogenesis, 2008. **29**(9): p. 1831–6.
28. Beezhold, K., et al., *miR-190-mediated downregulation of PHLPP contributes to arsenic-induced Akt activation and carcinogenesis.* Toxicol Sci, 2011. **123**(2): p. 411–20.
29. Koturbash, I., et al., *Epigenetic mechanisms of mouse interstrain variability in genotoxicity of the environmental toxicant 1,3-butadiene.* Toxicol Sci, 2011. **122**(2): p. 448–56.
30. Li, S., et al., *Developmental exposure to diethylstilbestrol elicits demethylation of estrogen-responsive lactoferrin gene in mouse uterus.* Cancer Res, 1997. **57**(19): p. 4356–9.
31. Tang, W.Y., et al., *Persistent hypomethylation in the promoter of nucleosomal binding protein 1 (Nsbp1) correlates with overexpression of Nsbp1 in mouse uteri neonatally exposed to diethylstilbestrol or genistein.* Endocrinology, 2008. **149**(12): p. 5922–31.
32. Li, S., et al., *Neonatal diethylstilbestrol exposure induces persistent elevation of c-fos expression and hypomethylation in its exon-4 in mouse uterus.* Mol Carcinog, 2003. **38**(2): p. 78–84.
33. Sato, K., et al., *Neonatal exposure to diethylstilbestrol alters expression of DNA methyltransferases and methylation of genomic DNA in the mouse uterus.* Endocr J, 2009. **56**(1): p. 131–9.
34. Watkeys, O.J., et al., *Glucocorticoid receptor gene (NR3C1) DNA methylation in association with trauma, psychopathology, transcript expression, or genotypic variation: A systematic review.* Neurosci Biobehav Rev, 2018. **95**: p. 85–122.
35. Murgatroyd, C., et al., *Dynamic DNA methylation programs persistent adverse effects of early-life stress.* Nat Neurosci, 2009. **12**(12): p. 1559–66.
36. Jeanneteau, F., et al., *The Stress-Induced Transcription Factor NR4A1 Adjusts Mitochondrial Function and Synapse Number in Prefrontal Cortex.* J Neurosci, 2018. **38**(6): p. 1335–1350.
37. Siuda, D., et al., *Social isolation-induced epigenetic changes in midbrain of adult mice.* J Physiol Pharmacol, 2014. **65**(2): p. 247–55.
38. Weaver, I.C., et al., *Epigenetic programming by maternal behavior.* Nat Neurosci, 2004. **7**(8): p. 847–54.
39. Murgatroyd, C. and D. Spengler, *Polycomb binding precedes early-life stress responsive DNA methylation at the Avp enhancer.* PLoS One, 2014. **9**(3): p. e90277.
40. Elliott, E., et al., *Resilience to social stress coincides with functional DNA methylation of the Crf gene in adult mice.* Nat Neurosci, 2010. **13**(11): p. 1351–3.
41. Franklin, T.B., et al., *Epigenetic transmission of the impact of early stress across generations.* Biol Psychiatry, 2010. **68**(5): p. 408–15.
42. Zhang, T.Y., et al., *Maternal care and DNA methylation of a glutamic acid decarboxylase 1 promoter in rat hippocampus.* J Neurosci, 2010. **30**(39): p. 13130–7.
43. Roth, T.L., et al., *Lasting epigenetic influence of early-life adversity on the BDNF gene.* Biol Psychiatry, 2009. **65**(9): p. 760–9.
44. Franklin, T.B., et al., *Influence of early stress on social abilities and serotonergic functions across generations in mice.* PLoS One, 2011. **6**(7): p. e21842.
45. Morris, M.J. and L.M. Monteggia, *Role of DNA methylation and the DNA methyltransferases in learning and memory.* Dialogues Clin Neurosci, 2014. **16**(3): p. 359–71.

46. McGowan, P.O., et al., *Epigenetic regulation of the glucocorticoid receptor in human brain associates with childhood abuse.* Nat Neurosci, 2009. **12**(3): p. 342–8.
47. Proal, A.D., I.A. Lindseth, and T.G. Marshall, *Microbe-microbe and host-microbe interactions drive microbiome dysbiosis and inflammatory processes.* Discov Med, 2017. **23**(124): p. 51–60.
48. Lane, N., *The unseen world: reflections on Leeuwenhoek (1677) 'Concerning little animals'.* Philos Trans R Soc Lond B Biol Sci, 2015. **370**(1666).
49. Integrative, H.M.P.R.N.C., *The Integrative Human Microbiome Project.* Nature, 2019. **569**(7758): p. 641–648.
50. Radaic, A. and Y.L. Kapila, *The oralome and its dysbiosis: New insights into oral microbiome-host interactions.* Comput Struct Biotechnol J, 2021. **19**: p. 1335–1360.
51. Pride, D.T., et al., *Evidence of a robust resident bacteriophage population revealed through analysis of the human salivary virome.* ISME J, 2012. **6**(5): p. 915–26.
52. Ly, M., et al., *Altered oral viral ecology in association with periodontal disease.* mBio, 2014. **5**(3): p. e01133–14.
53. Perez-Brocal, V. and A. Moya, *The analysis of the oral DNA virome reveals which viruses are widespread and rare among healthy young adults in Valencia (Spain).* PLoS One, 2018. **13**(2): p. e0191867.
54. He, J., et al., *The oral microbiome diversity and its relation to human diseases.* Folia Microbiol (Praha), 2015. **60**(1): p. 69–80.
55. Segata, N., et al., *Composition of the adult digestive tract bacterial microbiome based on seven mouth surfaces, tonsils, throat and stool samples.* Genome Biol, 2012. **13**(6): p. R42.
56. Urbaniak, C., et al., *Microbiota of human breast tissue.* Appl Environ Microbiol, 2014. **80**(10): p. 3007–14.
57. Costantini, L., et al., *Characterization of human breast tissue microbiota from core needle biopsies through the analysis of multi hypervariable 16S-rRNA gene regions.* Sci Rep, 2018. **8**(1): p. 16893.
58. Rinninella, E., et al., *What is the Healthy Gut Microbiota Composition? A Changing Ecosystem across Age, Environment, Diet, and Diseases.* Microorganisms, 2019. **7**(1).
59. Hallen-Adams, H.E. and M.J. Suhr, *Fungi in the healthy human gastrointestinal tract.* Virulence, 2017. **8**(3): p. 352–358.
60. Shah, S.A., et al., *Expanding known viral diversity in the healthy infant gut.* Nat Microbiol, 2023. **8**(5): p. 986–998.
61. Galland, L., *The gut microbiome and the brain.* J Med Food, 2014. **17**(12): p. 1261–72.
62. Gao, K., et al., *Tryptophan Metabolism: A Link Between the Gut Microbiota and Brain.* Adv Nutr, 2020. **11**(3): p. 709–723.
63. Varian, B.J., et al., *Microbial lysate upregulates host oxytocin.* Brain Behav Immun, 2017. **61**: p. 36–49.
64. Bravo, J.A., et al., *Ingestion of Lactobacillus strain regulates emotional behavior and central GABA receptor expression in a mouse via the vagus nerve.* Proc Natl Acad Sci U S A, 2011. **108**(38): p. 16050–5.
65. Han, Y., et al., *Vagus Nerve and Underlying Impact on the Gut Microbiota-Brain Axis in Behavior and Neurodegenerative Diseases.* J Inflamm Res, 2022. **15**: p. 6213–6230.
66. Zou, R., et al., *Changes in the Gut Microbiota of Children with Autism Spectrum Disorder.* Autism Res, 2020. **13**(9): p. 1614–1625.
67. Chen, X., et al., *The Female Vaginal Microbiome in Health and Bacterial Vaginosis.* Front Cell Infect Microbiol, 2021. **11**: p. 631972.
68. Giannella, L., et al., *Microbiome Changes in Pregnancy Disorders.* Antioxidants (Basel), 2023. **12**(2).
69. Feng, Y., et al., *Breast cancer development and progression: Risk factors, cancer stem cells, signaling pathways, genomics, and molecular pathogenesis.* Genes Dis, 2018. **5**(2): p. 77–106.
70. Lukasiewicz, S., et al., *Breast Cancer-Epidemiology, Risk Factors, Classification, Prognostic Markers, and Current Treatment Strategies-An Updated Review.* Cancers (Basel), 2021. **13**(17).
71. Sampsell, K., D. Hao, and R.A. Reimer, *The Gut Microbiota: A Potential Gateway to Improved Health Outcomes in Breast Cancer Treatment and Survivorship.* Int J Mol Sci, 2020. **21**(23).
72. Routy, B., et al., *Gut microbiome influences efficacy of PD-1-based immunotherapy against epithelial tumors.* Science, 2018. **359**(6371): p. 91–97.
73. Vetizou, M., et al., *Anticancer immunotherapy by CTLA-4 blockade relies on the gut microbiota.* Science, 2015. **350**(6264): p. 1079–84.
74. Mohseni, A.H., et al., *Potential links between the microbiota and T cell immunity determine the tumor cell fate.* Cell Death Dis, 2023. **14**(2): p. 154.
75. Fontenot, J.D., M.A. Gavin, and A.Y. Rudensky, *Foxp3 programs the development and function of CD4+CD25+ regulatory T cells.* Nat Immunol, 2003. **4**(4): p. 330–6.
76. Plitas, G., et al., *Regulatory T Cells Exhibit Distinct Features in Human Breast Cancer.* Immunity, 2016. **45**(5): p. 1122–1134.
77. Koshida, K., et al., *Dysfunction of Foxp3(+) Regulatory T Cells Induces Dysbiosis of Gut Microbiota via Aberrant Binding of Immunoglobulins to Microbes in the Intestinal Lumen.* Int J Mol Sci, 2023. **24**(10).
78. Parida, S., et al., *A Procarcinogenic Colon Microbe Promotes Breast Tumorigenesis and Metastatic Progression and Concomitantly Activates Notch*

*and beta-Catenin Axes*. Cancer Discov, 2021. **11**(5): p. 1138–1157.
79. Yang, L., et al., *Intratumoral microbiota: roles in cancer initiation, development and therapeutic efficacy*. Signal Transduct Target Ther, 2023. **8**(1): p. 35.
80. Rubinstein, M.R., et al., *Fusobacterium nucleatum promotes colorectal carcinogenesis by modulating E-cadherin/beta-catenin signaling via its FadA adhesin*. Cell Host Microbe, 2013. **14**(2): p. 195–206.
81. Lu, R., et al., *Enteric bacterial protein AvrA promotes colonic tumorigenesis and activates colonic beta-catenin signaling pathway*. Oncogenesis, 2014. **3**(6): p. e105.
82. Hatakeyama, M., *Helicobacter pylori CagA and gastric cancer: a paradigm for hit-and-run carcinogenesis*. Cell Host Microbe, 2014. **15**(3): p. 306–16.
83. Viswanathan, S., et al., *Role of Gut Microbiota in Breast Cancer and Drug Resistance*. Pathogens, 2023. **12**(3).
84. Yang, P., et al., *Comparison of the Gut Microbiota in Patients with Benign and Malignant Breast Tumors: A Pilot Study*. Evol Bioinform Online, 2021. **17**: p. 11769343211057573.
85. Desalegn, Z., et al., *Human Breast Tissue Microbiota Reveals Unique Microbial Signatures that Correlate with Prognostic Features in Adult Ethiopian Women with Breast Cancer*. Cancers (Basel), 2023. **15**(19).
86. Bloise, N., et al., *Targeting the "Sweet Side" of Tumor with Glycan-Binding Molecules Conjugated-Nanoparticles: Implications in Cancer Therapy and Diagnosis*. Nanomaterials (Basel), 2021. **11**(2).
87. Wu, H., S. Ganguly, and T.O. Tollefsbol, *Modulating Microbiota as a New Strategy for Breast Cancer Prevention and Treatment*. Microorganisms, 2022. **10**(9).
88. Bobin-Dubigeon, C., et al., *Faecal Microbiota Composition Varies between Patients with Breast Cancer and Healthy Women: A Comparative Case-Control Study*. Nutrients, 2021. **13**(8).
89. Plaza-Diaz, J. and A.I. Alvarez-Mercado, *The Interplay between Microbiota and Chemotherapy-Derived Metabolites in Breast Cancer*. Metabolites, 2023. **13**(6).
90. Al-Qadami, G.H., et al., *Gut Microbiota-Derived Short-Chain Fatty Acids: Impact on Cancer Treatment Response and Toxicities*. Microorganisms, 2022. **10**(10).
91. Banerjee, S., et al., *Prognostic correlations with the microbiome of breast cancer subtypes*. Cell Death Dis, 2021. **12**(9): p. 831.
92. Csendes, D., et al., *Gastrointestinal Microbiota and Breast Cancer Chemotherapy Interactions: A Systematic Review*. Cureus, 2022. **14**(11): p. e31648.
93. Dieleman, S., et al., *Exploring the Potential of Breast Microbiota as Biomarker for Breast Cancer and Therapeutic Response*. Am J Pathol, 2021. **191**(6): p. 968–982.
94. Hieken, T.J., et al., *The Microbiome of Aseptically Collected Human Breast Tissue in Benign and Malignant Disease*. Sci Rep, 2016. **6**: p. 30751.
95. Mousa, W.K., *The microbiome-product colibactin hits unique cellular targets mediating host-microbe interaction*. Front Pharmacol, 2022. **13**: p. 958012.
96. Shively, C.A., et al., *Consumption of Mediterranean versus Western Diet Leads to Distinct Mammary Gland Microbiome Populations*. Cell Rep, 2018. **25**(1): p. 47–56 e3.
97. Swales, K.E., et al., *The farnesoid X receptor is expressed in breast cancer and regulates apoptosis and aromatase expression*. Cancer Res, 2006. **66**(20): p. 10120–6.
98. Marin-Tello, C., et al., *Epigenetic regulation by metabolites from the gut microbiome*. Benef Microbes, 2022. **13**(6): p. 437–444.
99. Davie, J.R., *Inhibition of histone deacetylase activity by butyrate*. J Nutr, 2003. **133**(7 Suppl): p. 2485S-2493S.
100. Bishop, K.S., H. Xu, and G. Marlow, *Epigenetic Regulation of Gene Expression Induced by Butyrate in Colorectal Cancer: Involvement of MicroRNA*. Genet Epigenet, 2017. **9**: p. 1179237X17729900.
101. Zgouras, D., et al., *Butyrate impairs intestinal tumor cell-induced angiogenesis by inhibiting HIF-1alpha nuclear translocation*. Biochem Biophys Res Commun, 2003. **300**(4): p. 832–8.
102. Shenderov, B.A., *Metabiotics: novel idea or natural development of probiotic conception*. Microb Ecol Health Dis, 2013. **24**.
103. Ren, Z., et al., *NAD(+) and its possible role in gut microbiota: Insights on the mechanisms by which gut microbes influence host metabolism*. Anim Nutr, 2022. **10**: p. 360–371.
104. Wu, S.E., et al., *Microbiota-derived metabolite promotes HDAC3 activity in the gut*. Nature, 2020. **586**(7827): p. 108–112.
105. Meeran, S.M., S.N. Patel, and T.O. Tollefsbol, *Sulforaphane causes epigenetic repression of hTERT expression in human breast cancer cell lines*. PLoS One, 2010. **5**(7): p. e11457.
106. Shin, J.M., et al., *Biomedical applications of nisin*. J Appl Microbiol, 2016. **120**(6): p. 1449–65.
107. Laborda-Illanes, A., et al., *Breast and Gut Microbiota Action Mechanisms in Breast Cancer Pathogenesis and Treatment*. Cancers (Basel), 2020. **12**(9).
108. Li, L., et al., *Histone deacetylase inhibitor sodium butyrate suppresses DNA double strand break repair induced by etoposide more effectively in MCF-7 cells than in HEK293 cells*. BMC Biochem, 2015. **16**: p. 2.
109. Abe, M. and D.W. Kufe, *Effect of sodium butyrate on human breast carcinoma (MCF-7) cellular pro-*

*liferation, morphology, and CEA production*. Breast Cancer Res Treat, 1984. **4**(4): p. 269–74.
110. Chen, J., K.N. Zhao, and L. Vitetta, *Effects of Intestinal Microbial(-)Elaborated Butyrate on Oncogenic Signaling Pathways*. Nutrients, 2019. **11**(5).
111. Truong, D.T., et al., *MetaPhlAn2 for enhanced metagenomic taxonomic profiling*. Nat Methods, 2015. **12**(10): p. 902–3.
112. Wood, D.E. and S.L. Salzberg, *Kraken: ultrafast metagenomic sequence classification using exact alignments*. Genome Biol, 2014. **15**(3): p. R46.
113. Shakya, M., C.C. Lo, and P.S.G. Chain, *Advances and Challenges in Metatranscriptomic Analysis*. Front Genet, 2019. **10**: p. 904.
114. Jariyasopit, N. and S. Khoomrung, *Mass spectrometry-based analysis of gut microbial metabolites of aromatic amino acids*. Comput Struct Biotechnol J, 2023. **21**: p. 4777–4789.

# 5 Facilitates Chromatin Transcription in Breast and Other Cancers

Priyanka Barman and Sukesh R. Bhaumik

**Abstract**

Eukaryotic genome is packaged into chromatin. Thus, transcription takes place in the context of chromatin that is an array of nucleosomes. Nucleosome poses a barrier for the gene regulatory factors to access DNA for transcription to occur. Fortunately, eukaryotic cells have evolved mechanisms of nucleosomal disassembly and reassembly for transcription through chromatin. Such nucleosomal alteration in controlling transcription is governed by a heterodimeric chromatin remodeling factor, FACT (facilitates chromatin transcription), which is evolutionarily conserved from yeast to humans. FACT facilitates chromatin disassembly at the promoter and reassembly at the open reading frame. Such chromatin regulatory functions of FACT promote transcription. Likewise, other DNA transacting processes such as DNA replication and repair are also regulated by FACT via modulation of chromatin dynamics. Intriguingly, FACT is found to be upregulated in breast and other cancers with oncogenic potential. Thus, FACT and/or its upstream regulatory pathways/factors can be employed for cancer prognosis and targeted for an effective cancer therapy. Further, FACT is found to be downregulated and/or mutated in various cancers including breast cancer. Here, we describe FACT and its involvement in breast and other cancers with prognostic and targeted therapeutic implications.

**Keywords**

FACT (facilitates chromatin transcription) · Chromatin · Transcription · RNA polymerase II · Cancer · cBioPortal

## 5.1 Introduction

Eukaryotic genome is compacted to fit in the limited space within the nucleus. Such compaction is mediated via the formation of chromatin that is an array of nucleosomes. Each nucleosome contains ~146 base pairs of DNA wrapping around a histone octamer consisting of a histone H3–H4 tetramer and two histone H2A–H2B dimers [1]. Such organization of eukaryotic genome often blocks the access of DNA-binding factors to DNA. Thus, chromatin structure and associated chromatin regulatory factors control epigenetically transcription and other DNA-transacting processes such as DNA replication and repair [2, 3]. Fortunately, eukaryotic genome undergoes dynamic structural changes that include nucleo-

P. Barman · S. R. Bhaumik (✉)
Department of Biochemistry and Molecular Biology, Southern Illinois University School of Medicine, Carbondale, IL, USA
e-mail: sbhaumik@siumed.edu



S. Bhatnagar (ed.), *Epigenetic Mechanisms in Breast Cancer Therapy and Resistance*, Advances in Experimental Medicine and Biology 1465, https://doi.org/10.1007/978-3-031-66686-5_5

somal disassembly and reassembly to allow and control transcription and other DNA-transacting processes [4–11]. Such dynamic nucleosomal disassembly and reassembly are governed by chaperones that interact with histones in an ATP-independent fashion [12–15]. One such histone chaperone is an evolutionarily conserved heterodimeric FACT (facilitates chromatin transcription) [12–18]. FACT was originally identified by Reinberg and colleagues [19] as an essential factor to promote transcription on a chromatinized, but not on non-chromatinized, DNA template and was later found to be involved in regulating other DNA-transacting processes [13–15, 17]. Thus, FACT is an essential factor for normal cellular functions. Hence, its misregulation, malfunction, or mutation can be associated with cellular pathologies/diseases. Indeed, FACT is found to be upregulated, downregulated, or mutated in various cancers including breast cancer ([20–23]; cBioPortal). Therefore, FACT and/or the regulatory factors/mechanisms lying upstream of FACT can be attractive targets for cancer therapy. Further, altered levels of FACT and/or FACT regulatory factors can be prognostic cancer biomarkers. Here, we describe FACT and its involvement in breast and other cancers with implications in pathogenesis and targeted therapeutic development.

## 5.2 FACT

FACT is a heterodimer of SPT16 (suppressor of Ty 16) and SSRP1 (structure specific recognition protein 1) in higher eukaryotes [19, 24, 25]. Likewise, in lower eukaryote (e.g., yeast), FACT contains two subunits, Spt16 and Pob3 (polymerase one binding protein 3). Pob3 is a yeast homologue of SSRP1, but unlike SSRP1, it does not have a HMGB (high mobility group box)-like domain. HMGB-like domain or HMGB protein is involved in interaction with nucleosome. Thus, for interaction with nucleosome, yeast FACT depends on an HMGB protein, Nhp6 [26].

Both subunits of FACT have conserved domains. SPT16 has four domains, namely, NTD (N-terminal domain), DD (dimerization domain), MD (middle domain), and CTD (C-terminal domain), while SSRP1 has NTD, MD, IDD (intrinsically disordered domain), HMGB, and intrinsically disordered CTD (CID) (Fig. 5.1). The SPT16 DD interacts with the NTD of SSRP1 to form the FACT complex (Fig. 5.1). Other domains of FACT are involved in interactions with histones. For example, SPT16 NTD interacts with the N-terminal tails of histones H3 and H4, and its CTD binds to histone H2A-H2B dimer [27–36]. SSRP1 MD interacts with histone H3-H4 tetramer [17]. Further, FACT contacts with nucleosomal DNA [17, 37–41]. SSRP1 HMGB plays an important role in such interaction [17, 37–41]. Through these interactions, FACT regulates chromatin. FACT is involved in both chromatin/nucleosomal disassembly and reassembly to overcome and maintain nucleosomal barrier for DNA-transacting processes to occur. During transcription, FACT promotes nucleosomal disassembly at the promoter and reassembly at the open reading frame (ORF) in the wake of elongating RNA polymerase II [42–51]. FACT's role in promoting chromatin disassembly at the promoter favors the formation of the pre-initiation complex for transcription initiation. FACT-mediated chromatin reassembly at the ORF behind elongating RNA polymerase II suppresses cryptic transcription (that is associated with ageing; [52–56]) and promotes transcription elongation. FACT's function in chromatin reassembly is dependent on histone H2B mono-ubiquitylation (non-proteolytic or that is not targeted for 26S proteasomal degradation) which occurs via the sequential actions of E1 activating enzymes, E2 conjugase, and E3 ligase [57–60]. In addition, histone H2B ubiquitylation is controlled by DUB (de-ubiquitinase or ubiquitin protease) that reverses ubiquitylation [60–62]. Further, histone H2B ubiquitylation is found to be regulated by a key transcription factor, namely, Paf1C (polymerase-associated factor 1 complex) [63–66]. Thus, FACT and Paf1C are functionally linked to regulate chromatin reassembly via histone H2B ubiquitylation. Further, Paf1C regulates transcription by facilitating histone H3 K4 (lysine 4) methylation [67, 68]. Histone H3 K4 methylation is also regulated by

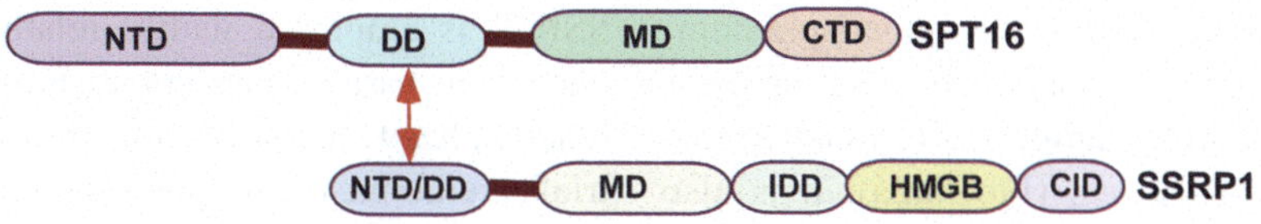


**Fig. 5.1** Schematic diagram showing different domains of the SPT16 and SSRP1 subunits of FACT in humans. Double arrowheaded red line indicates interaction. *NTD* N-terminal domain, *DD* dimerization domain, *MD* middle domain, *CTD* C-terminal domain, *IDD* intrinsically disordered domain, *HMGB* high mobility group box, *CID* intrinsically disordered CTD

histone H2B ubiquitylation via trans-tail cross-talk [69–74]. Thus, Paf1C is involved in regulating both histone H2B ubiquitylation and histone H3 K4 methylation that are associated with gene activation [67, 68]. Further, as described above, via its regulatory function of histone H2B ubiquitylation, Paf1C is also involved in controlling chromatin reassembly at the ORF in the wake of elongating RNA polymerase II.

FACT-mediated alterations of chromatin/nucleosome involve the disruption of the interaction between histone H2A-H2B dimer and DNA [27]. Such disruption occurs via the competitive binding of the negatively charged and intrinsically disordered CTD of SPT16 with nucleosomal histone H2A-H2B dimer and allows the unraveling and accessibility of nucleosomal DNA. However, the nucleosomal subunits stay bound to FACT so that it can be reassembled following completion of the DNA transacting process. In addition to regulate chromatin and chromatin-associated processes, FACT is also involved in controlling other cellular events. For example, FACT is found to interact with mRNA export factor for nuclear export [75, 76]. It resolves R-loop (that is formed via hybridization of nascent mRNA with DNA behind elongating RNA polymerase II) during transcription elongation so that transcription-replication collision and hyper-recombination can be avoided to maintain genomic stability [77]. It is also involved in facilitating the release of paused-RNA polymerase II from the promoter proximal site for productive transcription elongation via its interaction with mRNA capping machinery and Paf1C [78]. Further, it interacts with E3 ubiquitin ligase, Psh1, that ubiquitylates centromeric histone H3 for proteasomal degradation so that active gene is devoid of centromeric histone H3 [79]. In addition to interacting with Psh1, it also binds to another E3 ubiquitin ligase (Rtt101 in yeast and DCAF14-CRL4 in humans) that ubiquitylates its SPT16 subunit, which does not undergo proteasomal degradation but rather regulates DNA replication [80, 81]. Further, SPT16 interacts with another E3 ubiquitin ligase (San1 in yeast, but yet unknown in humans) that undergoes ubiquitylation for 26S proteasomal degradation [82]. Such ubiquitylation and 26S proteasomal degradation of SPT16 orchestrate transcription and DNA repair [82]. Thus, FACT undergoes both proteolytic and non-proteolytic ubiquitylation to regulate gene expression and genome maintenance. In addition to be regulated by ubiquitylation and/or proteasomal degradation, FACT is also controlled by other posttranslational modifications such as phosphorylation and hydroxylation for its functions [83, 84]. Further, FACT participates in the formation of heterochromatin and centromere [85–92]. Thus, FACT functions in various cellular processes, and hence, its malfunction, misregulation, or mutation is associated with cellular pathologies and cancer ([20–23, 93]; cBioPortal). We describe below the involvement of FACT in various cancers.

## 5.3 FACT in Various Cancers

To find out the involvement of FACT in cancer, we analyzed SPT16 and SSRP1 in various cancer patient samples available in cBioPortal [94–97]. We looked for the amplifications, deletions, mutations, and structural variants (or gene fusions) of SPT16 and SSRP1 in 67,030 patients (or 70,655 samples) from 217 studies in cBioPortal, using minimum frequency and sample number of 0.1% and 3, respectively. Our analysis revealed that SPT16 is amplified or mutated in a large number of cancers (Fig. 5.2a). It is also found to be deleted in several cancers (Fig. 5.2a). Further, SPT16 is fused with BRMS1 (breast cancer metastasis suppressor 1) in undifferentiated pleomorphic sarcoma/malignant fibrous histiocytoma/high-grade spindle cell sarcoma (cBioPortal). There is a total of 447 mutations (including 15 duplications in cancer patients with multiple samples) in SPT16 in various cancers (Fig. 5.2b). These mutations are spread over the whole SPT16 and include 355 missense, 56 truncation, 30 splice, 5 inframe, and 1 fusion (structural variant) (Fig. 5.2b). Many of these mutations occurred on the K residues of SPT16, which could affect SPT16 function and/or regulation, as K is a prime site for a number of post-translational modifications including ubiquitylation and/or proteasomal degradation. These mutations of the SPT16 K residues are associated with various cancers (Table 5.1). In addition to having mutations, SPT16 is also amplified with a relatively high frequency in a large number of cancers such as lung, ovarian, breast, pancreatic, head, and neck cancers (Fig. 5.2a). Further, SPT16 is also found to be deleted with a relatively low frequency in some cancers such as soft tissue myoepithelial carcinoma, meningioma and renal non-clear cell carcinoma (Fig. 5.2a). Taken together, SPT16 mutation, fusion, amplification, and deletion are associated with various cancers.

Similar to SPT16, SSRP1 is also amplified, deleted, or mutated in various cancers (Fig. 5.2c). Further, like SPT16, SSRP1 is fused with GANAB (*gene* for neutral alpha-glucosidase AB precursor) and MED19 (mediator complex subunit 19) in serous ovarian cancer (cBioPortal). SSRP1 is found to have 278 mutations (including 4 duplicates in patients with multiple samples) in various cancer (cBioPortal). These mutations include 230 missense, 28 truncation, 15 splice, 3 inframe, and 2 fusion and are spread over whole SSRP1 (Fig. 5.2d). Similar to SPT16, many K residues of SSRP1 are mutated in various cancers (Table 5.2). In addition to having mutations, SSRP1 is amplified with a relatively high frequency in various cancers such as lung, breast, bladder, head, and neck cancers, uterine endometrial carcinoma, pleural mesothelioma, pheochromocytoma, and soft tissue sarcoma (Fig. 5.2c). Further, it is depleted with a relatively low frequency in some cancers such as pleural mesothelioma and breast and ovarian cancers (Fig. 5.2c), Thus, like SPT16, the SSRP1 subunit of FACT is also mutated, deleted, or amplified in various cancers (Fig. 5.2c, d; Table 5.2).

Both SPT16 and SSRP1 are amplified in a number of cancers such as breast, lung, ovarian, head, and neck cancers (Fig. 5.2a, c). Likewise, both subunits of FACT are deleted with low frequency in some cancers (e.g., ovarian and breast cancers) (Fig. 5.2a, c). Similarly, mutations in FACT subunits are commonly found in skin, breast, bladder, and colorectal cancers (Fig. 5.2a, c). Thus, both SPT16 and SSRP1 are found to be amplified, deleted, or mutated in a number of cancers (Fig. 5.2a, c). Intriguingly, FACT subunits are uniquely altered in certain cancers. For example, only SSRP1, but not SPT16, is amplified in uterine endometrial cancer, while only SPT16 is found to have mutations in this cancer (Fig. 5.2a, c). SPT16 is deleted in meningioma, but no alteration in SSRP1 is found in this cancer (Fig. 5.2a, c). Breast sarcoma is found to be associated with mutations in SPT16, but not in SSRP1 (Fig. 5.2a, c). Similarly, other cancers such as soft tissue myoepithelial carcinoma and cervical and vaginal cancers show differential alterations of SPT16 and SSRP1 (Fig. 5.2a, c). Thus, alterations of either or both FACT subunits are associated with cancers.

Since both components of FACT are altered in various cancers, and many of the mutations of FACT subunits are truncated, amplification, deletion, or mutation of FACT subunits would affect their mRNA or protein levels. Therefore, we looked into the mRNA and protein levels of SPT16 and SSRP1 in various cancer patient samples in UALCAN [98, 99] Web portal (The University of ALabama at Birmingham CANcer data analysis Portal). We find that mRNA and protein levels of SPT16 and SSRP1 are changed in cancer patients' samples (Fig. 5.3a–d). In

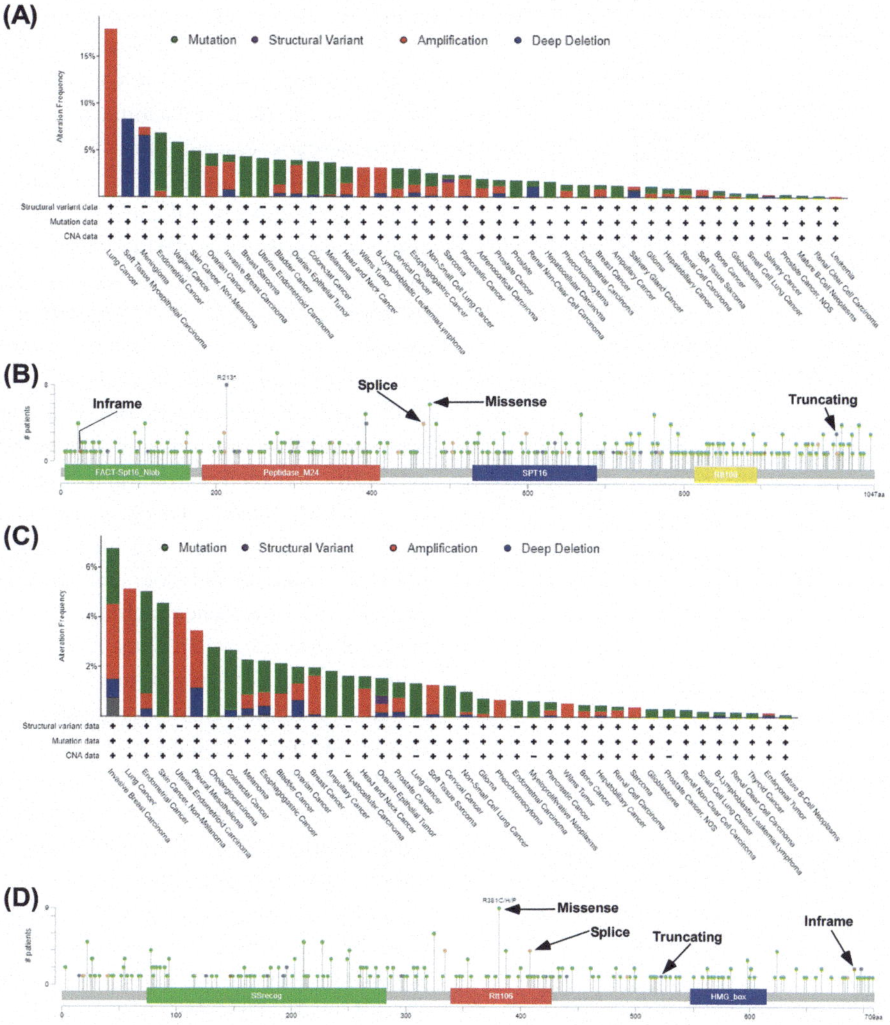


**Fig. 5.2** Alterations of SPT16 and SSRP1 in different cancer patient samples. (**a**) Cross-cancer analysis of copy number alterations (CNA) and mutations in *SUPT16H (gene for SPT16)* based on the patient samples in the cBioPortal database [217 studies that are manually curated including TCGA (the cancer genome atlas) and non-TCGA studies with no overlapping samples]. All patient samples (a total of 70,655 samples/67,030 patients from 217 studies) were analyzed in cBioPortal. The frequencies of 0.1% and above were plotted with a minimum sample number of 3. (**b**) Mutations at different residues/amino acids (aa) of *SUPT16H* in various cancer patient samples. These mutations include K (lysine) and other amino acids. There is a total of 447 mutations (including 15 duplicates in the cancer patients with multiple samples). These include 355 missense, 56 truncating, 5 inframe, 30 splice, and 1 fusion (structural variant). (**c**) Cross-cancer analysis of CNA and mutations in *SSRP1,* based on the patient samples in the cBioPortal database [217 studies that are manually curated including TCGA and non-TCGA studies with no overlapping samples]. All patient samples (a total of 70,655 samples/67,030 patients from 217 studies) were analyzed in cBioPortal. The frequencies of 0.1% and above were plotted with a minimum sample number of 3. (**d**) Mutations at different residues/amino acids of *SSRP1* in various cancer patient samples. These mutations include K and other amino acids. There is a total of 278 mutations (including 4 duplicates in the cancer patients with multiple samples). These include 230 missense, 28 truncating, 3 inframe, 15 splice, and 2 fusion (structural variant)

**Table 5.1** K (lysine) mutations of SPT16 in various cancer patients' samples

| Cancers | K mutations |
|---|---|
| Activated B-cell type | K126E |
| Lung squamous cell carcinoma | K792 (nonsense), K1019N |
| Stomach adenocarcinoma | K346E |
| Cutaneous melanoma | K883N |
| Colon adenocarcinoma | K626E, K392E, K393fs (frame shift) |
| Uterine endometrial carcinoma | K393fs, K389E, K1043Rfs, K732N, K696R |
| Intestine-type stomach adenocarcinoma | K1043Rfs |
| Head and neck squamous cell carcinoma | K732N, K392E |
| Prostate adenocarcinoma | K792Q |
| Bladder urothelial carcinoma | K132N, K392E |
| Lung adenocarcinoma | K792R |
| Colorectal adenocarcinoma | K392E, K393fs, K393R, K223Rfs |
| Ampullary carcinoma | K229Nfs |

**Table 5.2** K mutations of SSRP1 in various cancer patients' samples

| Cancers | K mutations |
|---|---|
| Endometrial cancer | K663Qfs |
| Cutaneous melanoma | K566N |
| Rectal adenocarcinoma | K650N |
| Colon adenocarcinoma | K228E, K33E |
| Lung adenocarcinoma | K296fs, K90T, K663R |
| Cutaneous squamous cell carcinoma | K461* (*, nonsense) |
| Serous ovarian cancer | K561M |
| Glioblastoma | K628N |
| Uterine endometrial carcinoma | K522Rfs, K630N, K426T, K530N |
| Prostate adenocarcinoma | K518R |
| Papillary renal cell carcinoma | K15* |
| Glioblastoma multiforme | K461N, K67N |
| Invasive breast cancer | K90R |

many cancers, protein and mRNA levels of SPT16 and SSRP1 follow the similar trends, i.e., increase or decrease of both mRNA and protein levels (Fig. 5.3a–d). Intriguingly, in some cancers, mRNA and protein levels of SPT16 do not follow the similar trends. For example, mRNA level of SPT16 decreases while its protein level increases in renal clear cell carcinoma (Fig. 5.3a, b). There is significant increase in SPT16 protein, but not mRNA level in several cancers (Fig. 5.3a, b). There is much more increase in SPT16 protein level in comparison to mRNA in several cancer patients such as glioblastoma and breast, lung, and pancreatic cancers (Fig. 5.3a, b). It is not clear why SPT16 mRNA and protein levels are not correlated in several cancers (Fig. 5.3a, b). It could be possible that SPT16 protein is regulated via ubiquitylation and proteasomal degradation in normal cells, and such regulation might be impaired in cancer cell, thereby contributing to increased protein level. Indeed, we found SPT16 ubiquitylation and proteasomal degradation in orchestrating transcription in human and mouse cells [20], and such regulation is impaired in cancer cells [20]. Further, we observed that increased level of SPT16 (that can be resulted by its impaired proteasomal degradation) changes the expression of the genes including the ones associated with cancer [20]. Thus, these results imply impairment of SPT16 ubiquitylation and proteasomal degradation in its upregulation in cancer. In addition to be regulated via ubiquitylation and proteasomal degradation, SPT16 is also regulated by other posttranslational modifications such as hydroxylation and phosphorylation [83, 84], which could be misregulated in cancer, leading to its alteration. Further, epigenetic, transcriptional, or posttranscriptional regulation of SPT16 might be changed in cancer cells, contributing to its altered levels. Therefore, future studies are warranted to understand FACT regulation at various steps of gene expression and its posttranslational modifications and determine what goes awry in cancer cells to decipher new mechanisms of cancer with targeted therapeutic potentials.

## 5.4 FACT in Breast Cancer

As described above, alterations of the FACT subunits are associated with breast cancer. To determine the involvement of FACT in different types of breast cancers, we analyzed the SPT16 and SSRP1 subunits of FACT in breast invasive cancer, breast invasive lobular cancer, breast invasive ductal cancer, breast invasive mixed mucinous cancer, and breast invasive mixed ductal and lob-

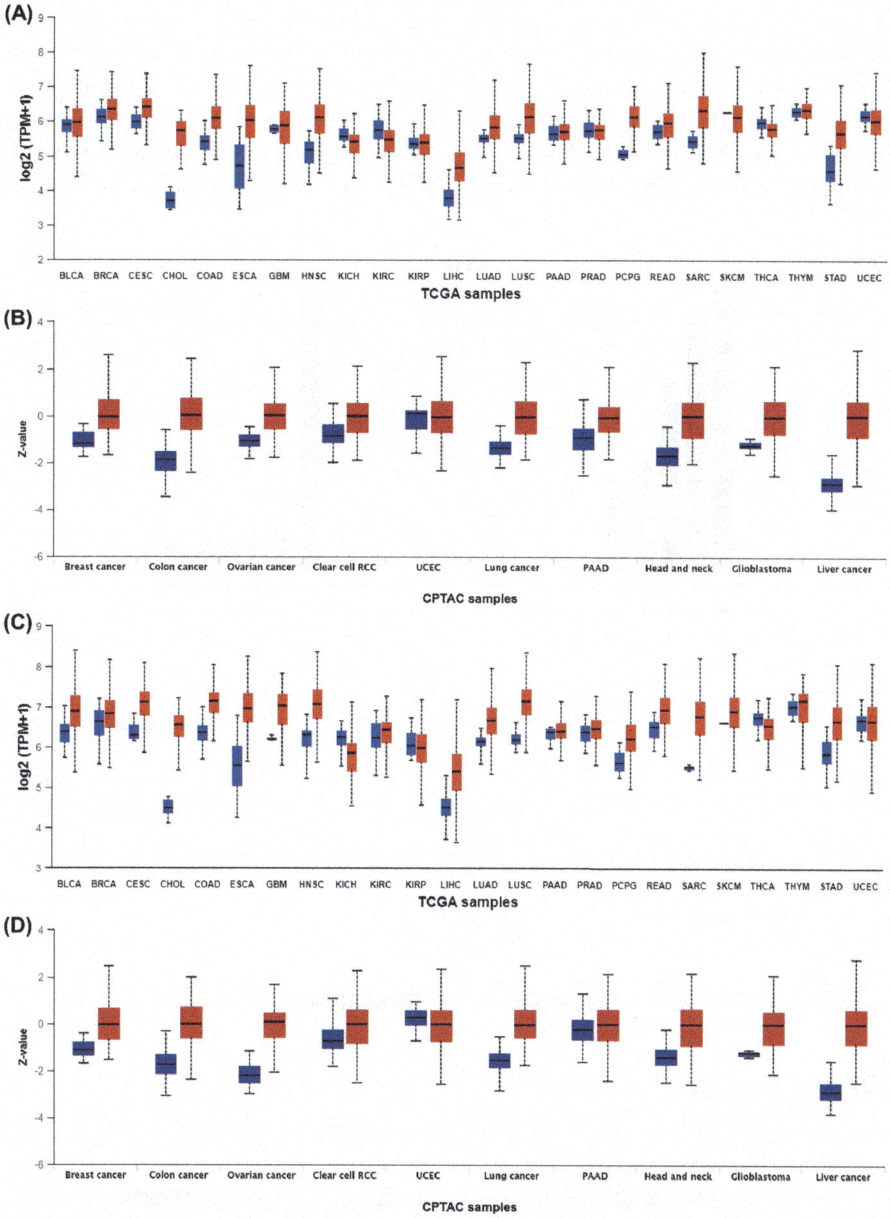


**Fig. 5.3** Protein and mRNA levels of SPT16 and SSRP1 in various cancer patient samples. (**a**) *SUPT16H* mRNA levels in various cancer patient samples based on TCGA in the UALCAN. TPM (transcripts per million) is a normalized number and is read as the number of RNA of a particular gene of interest out of one million RNA molecules. Red color represents cancer patient samples, and normal is referred by light blue color. *KIRC* kidney renal clear cell carcinoma, *BLCA* bladder urothelial carcinoma, *BRCA* breast invasive carcinoma, *CESC* cervical squamous cell carcinoma, *CHOL* cholangiocarcinoma, *COAD* colon adenocarcinoma, *ESCA* esophageal carcinoma, *GBM* glioblastoma multiforme, *HNSC* head and neck squamous cell carcinoma, *KICH* kidney chromophobe, *KIRP* kidney renal papillary cell carcinoma, *LIHC* liver hepatocellular carcinoma, *LUAD* lung adenocarcinoma, *LUSC* lung squamous cell carcinoma, *PAAD* pancreatic adenocarcinoma, *PRAD* prostate adenocarcinoma, *PCPG* pheochromocytoma and paraganglioma, *READ* rectal carcinoma, *SARC* sarcoma, *SKCM* skin cutaneous melanoma, *THCA* thyroid carcinoma, *THYM* thymoma, *STAD* stomach adenocarcinoma, *UCEC* uterine corpus endometrial carcinoma. (**b**) SPT16 protein levels in various cancer patient samples based on CPTAC (clinical proteomic tumor analysis consortium) in the UALCAN. The *Z* values for SPT16 in normal and cancer samples are plotted. Z value represents standard deviation from the median across samples for the given cancer type. RCC, renal cell carcinoma. (**c**) SSRP1 mRNA levels in various cancer patient samples based on TCGA in the UALCAN. (**d**) SSRP1 protein levels in various cancer patient samples based on CPTAC in the UALCAN

ular cancer using available breast cancer patient samples (8930 samples/8149 patients from 23 studies) in cBioPortal (with 0.1% and 3 minimum frequency and sample number, respectively) (Fig. 5.4a, b). We find that SSRP1 is mostly amplified in different types of breast cancers, while a relatively low frequency of its deletion and mutation is observed in some breast cancers (Fig. 5.4b). Likewise, mutation, amplification, and deletion are also observed for SPT16 in different breast cancer types (Fig. 5.4a). However, SSRP1 amplification is found to have a

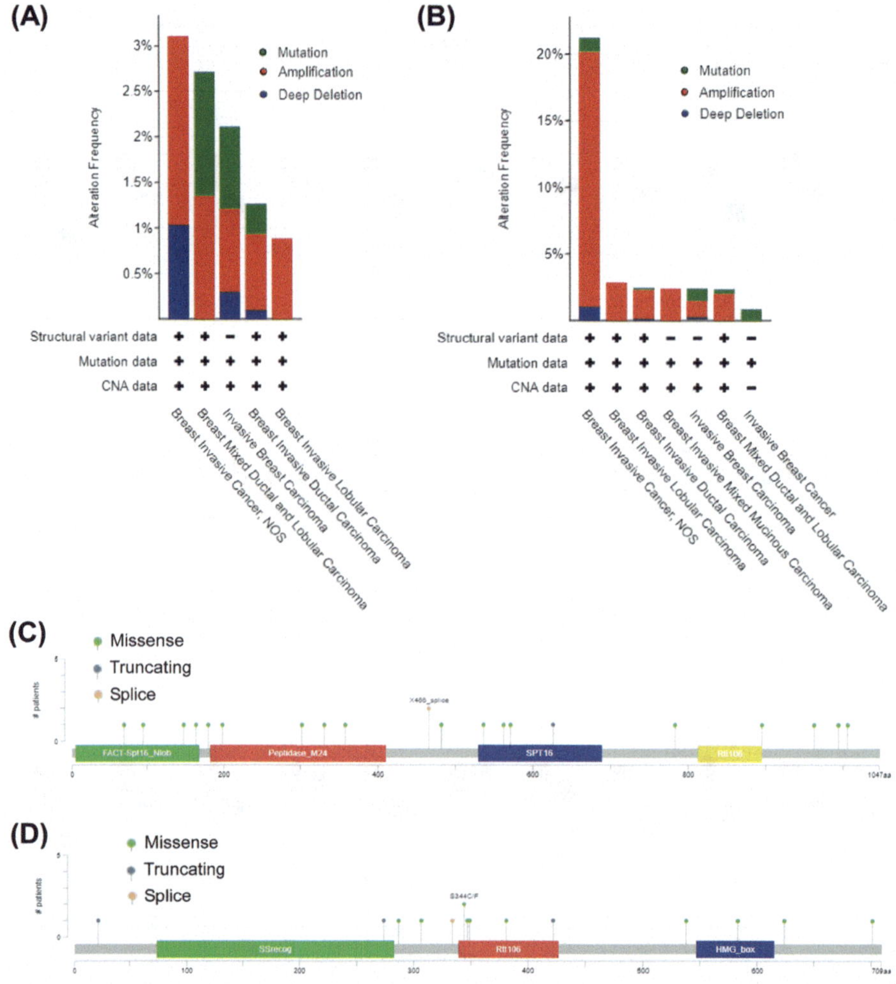


**Fig. 5.4** Alterations of SPT16 and SSRP1 in different breast cancer patient samples. (**a**) Cross-cancer analysis of CNA and mutations in *SSRP1,* based on the breast cancer patient samples in the cBioPortal database [23 studies that are manually curated including TCGA and non-TCGA studies with no overlapping samples]. All patient samples (a total of 8930 samples/8149 patients from 23 studies) were analyzed in cBioPortal. The frequencies of 0.1% and above were plotted with a minimum sample number of 3. (**b**) Cross-cancer analysis of CNA and mutations in *SSRP1,* based on the patient samples in the cBioPortal database [23 studies that are manually curated including TCGA and non-TCGA studies with no overlapping samples]. All patient samples (a total of 8930 samples/8149 patients from 23 studies) were analyzed in cBioPortal. The frequencies of 0.1% and above were plotted with a minimum sample number of 3. (**c**) Mutations at different residues/aa of SPT16 in various cancer patient samples. There is a total of 25 mutations (including 3 duplicates in the breast cancer patients with multiple samples). These include 21 missense, 2 truncating, and 2 splice mutations. (**d**) Mutations at different residues/aa of SSRP1 in various cancer patient samples. There is a total of 16 mutations (including 1 duplicate in the breast cancer patients with multiple samples). These include 12 missense, 3 truncating, and 1 splice mutations

relatively high frequency, while a relatively low frequency of SPT16 amplification is observed in breast invasive cancer (Fig. 5.4a, b). SPT16 and SSRP1 are found to have 25 and 16 mutations (including 3 and 1 duplicates in breast cancer patients with multiple samples), respectively, in different breast cancers (Fig. 5.4c, d). These mutations include missense (21 for SPT16 and 12 for SSRP1), truncation (2 for SPT16 and 3 for SSRP1), and splice (2 for SPT16 and 1 for SSRP1). However, mutations in K residues of SSRP1 and SPT16 are not found in breast cancer patient samples in cBioPortal.

Next, we looked into the mRNA and protein levels of SPT16 and SSRP1 in different breast cancer patient samples in UALCAN Web portal (Fig. 5.5a–d). We observe a small (or marginal) increase in SSRP1 mRNA level but a greater increase in its protein level in breast cancer patients' samples (Fig. 5.5c, d). Similarly, small increase in SPT16 mRNA but a greater increase in its protein level was observed in breast cancer patients' samples (Fig. 5.5a, b). Subsequently, we analyzed SPT16 mRNA and protein levels in different types (luminal, Her2-positive, and triple-negative breast cancers), stages (stage 1, 2, 3, or 4), ages, and races (Caucasian, African-American, and Asian) of breast cancer patients (Fig. 5.6a–h). We find that SPT16 protein levels are high in luminal, Her2-positive, and triple-negative breast cancers in comparison to normal samples (Fig. 5.6a, b). However, a relatively greater increase in SPT16 protein level was observed for the triple-negative breast cancer patients (Fig. 5.6b). While we find much increase in SPT16 protein levels in these breast cancers, very small or no increase in mRNA level was observed (Fig. 5.6a). Likewise, a very small or no increase in SPT16 mRNA level was observed in all stages of breast cancer (Fig. 5.6c) while a greater increase in SPT16 protein level was observed in all stages of breast cancers (Fig. 5.6d). Similarly, there is a greater increase in SPT16 protein level in all ages of breast cancer (Fig. 5.6f), but a small or no increase in SPT16 mRNA level was observed at different ages of breast cancer patients (Fig. 5.6e). Further, there is a greater increase in SPT16 protein level for all races (especially African-American), but very little (for Caucasians and

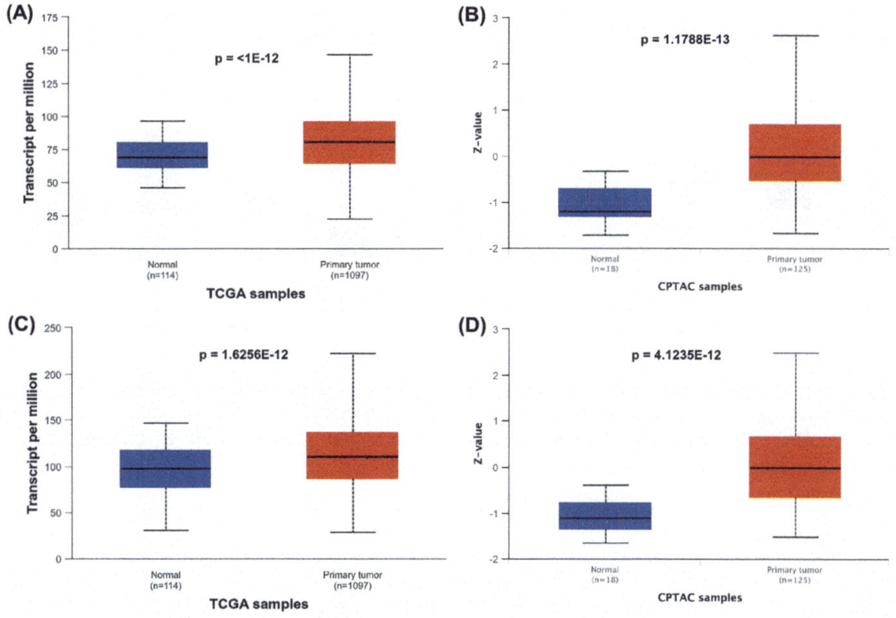


**Fig. 5.5** Analysis of SPT16 and SSRP1 mRNA and protein levels in the breast cancer samples in the UALCAN. (**a**) *SUPT16H* mRNA level in breast cancer. n, number of normal or tumor samples. (**b**) SPT16 protein level in breast cancer. (**c**) *SSRP1* mRNA level in breast cancer. (**d**) SSRP1 protein level in breast cancer.

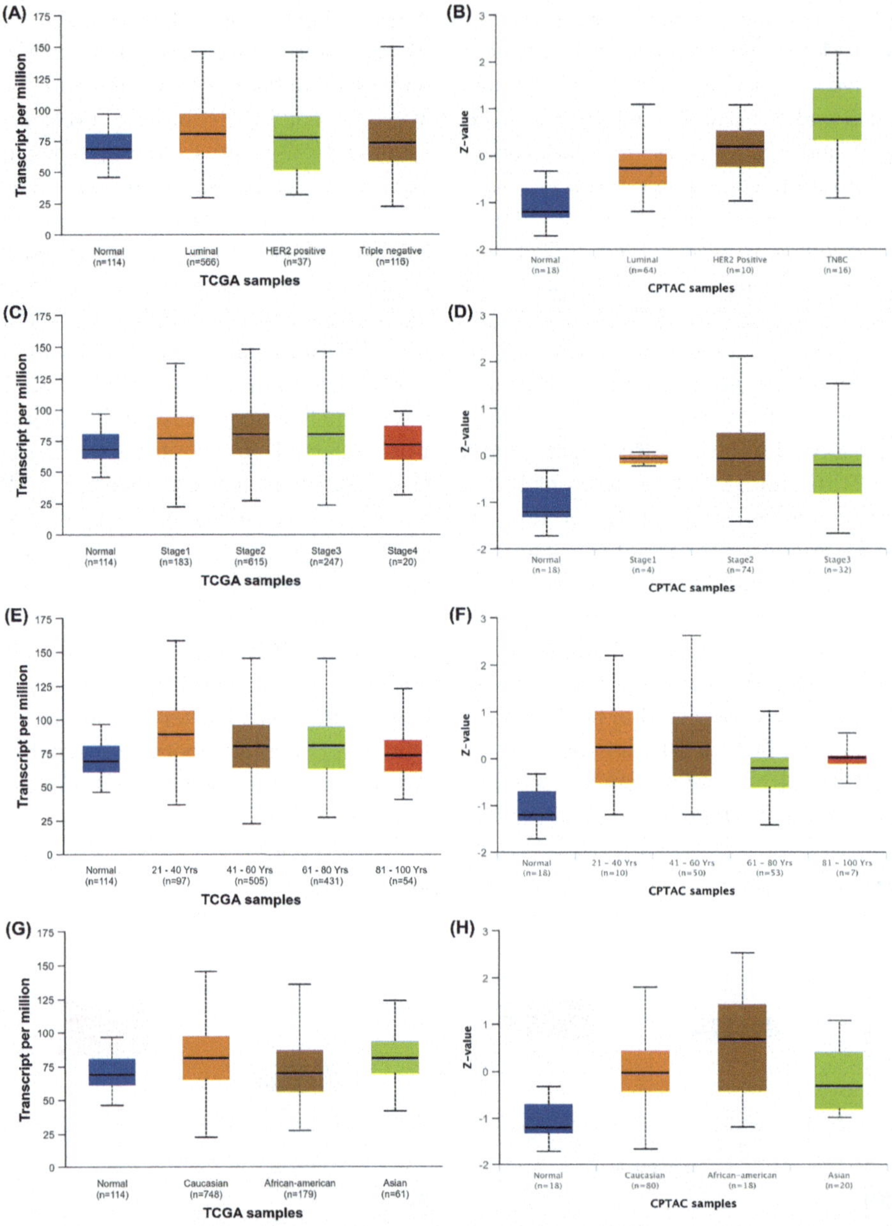


**Fig. 5.6** Analysis of SPT16 mRNA and protein levels in different types, stages, ages, and races of the breast cancer samples in the UALCAN. (**a** and **b**) *SUPT16H* mRNA and protein levels in different types of breast cancer. (**c** and **d**) *SUPT16H* mRNA and protein levels in different stages of breast cancer. (**e** and **f**) *SUPT16H* mRNA and protein levels in different ages of breast cancer. (**g** and **h**) *SUPT16H* mRNA and protein levels in different races of breast cancer.

Asians) or no increase (for African-American) in SPT16 mRNA levels was observed (Fig. 5.6g, h).

We next looked at SSRP1 mRNA and protein levels in different types, stages, ages, and races of breast cancer patients (Fig. 5.7a–h). We find no change in SSRP1 mRNA level in luminal and Her2 plus breast cancers, but triple-negative breast cancer patients show increase in SSRP1

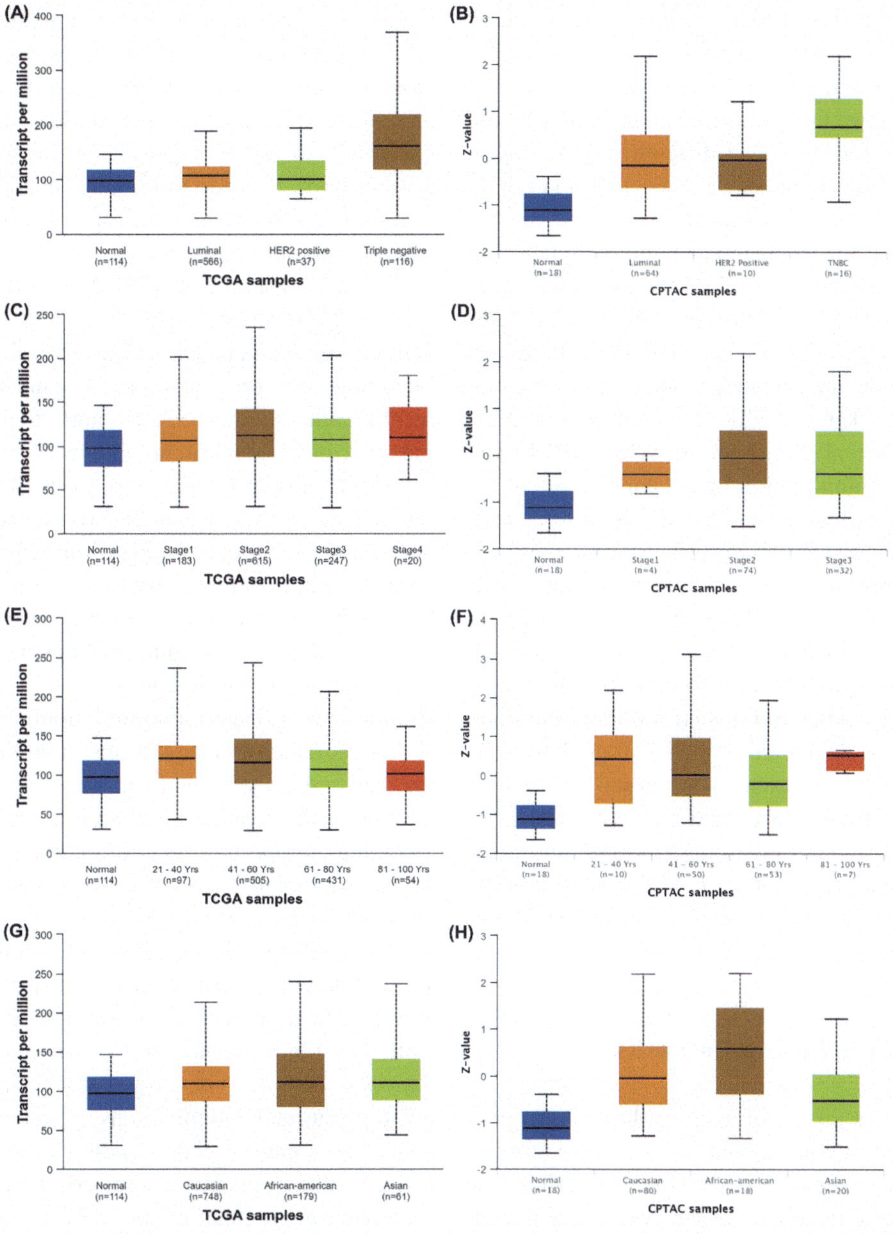


**Fig. 5.7** Analysis of SSRP1 mRNA and protein levels in different types, stages, ages, and races of the breast cancer samples in the UALCAN. (**a** and **b**) *SSRP1* mRNA and protein levels in different types of breast cancer. (**c** and **d**) *SSRP1* mRNA and protein levels in different stages of breast cancer. (**e** and **f**) *SSRP1* mRNA and protein levels in different ages of breast cancer. (**g** and **h**) *SSRP1* mRNA and protein levels in different races of breast cancer.

mRNA level (Fig. 5.7a). However, we observed increase in SSRP1 protein levels in all these breast cancer patients (Fig. 5.7b). There is a very small or no change in SSRP1 mRNA level in different ages of breast cancer patients (Fig. 5.7e) but rather much increase in SSRP1 protein levels (Fig. 5.7f). Likewise, we observed a very small or no change in SSRP1 mRNA level in different stages of breast cancer, but protein levels increased (Fig. 5.7c, d). Similarly, almost no change in SSRP1 mRNA level was observed in different races of breast cancer patients, but increased levels of SSRP1 protein were observed (Fig. 5.7g, h). Most increase was observed for African-Americans (Fig. 5.7h).

Overall, we find greater increase in SPT16 and SSRP1 protein levels in different types, stages, ages, and races of breast cancers but a small or no increase in their mRNA levels. Increase in mRNA level is not due to promoter hypomethylation as the gene promoters of SPT16 and SSRP1 are hypomethylated under normal and cancer states (Fig. 5.8a, b). Further, we find that alterations (that include amplifications, deletions, and mutations) of SPT16 and SSRP1 do not greatly affect overall survival (Fig. 5.9a, b). While SPT16 and SSRP1 are amplified in breast cancer patients, it is not clear why their mRNA levels show a small or no change in different types, stages, ages, and races of breast cancers, but their protein levels are greatly increased. This could be due to alteration of posttranslational and posttranscriptional regulation of SPT16 and SSRP1. In agreement, our recent studies implicated ubiquitylation and proteasomal degradation of SPT16 in altering its protein level in cancer as described above. However, future investigations are needed to deeply understand posttranscriptional and posttranslational regulation of SPT16 and SSRP1 with links to their altered mRNA/protein levels in breast cancer patients.

## 5.5 Concluding Remarks

FACT is an essential and evolutionarily conserved chromatin regulatory factor to control gene expression and genome integrity. Alteration of FACT and its functions is associated with cancer and other diseases. Likewise, FACT plays important roles in plants, and its deficiency shows various vegetative and reproductive defects [87, 100, 101]. As described above, FACT subunits are found to be amplified, deleted, or mutated in breast and other cancers. However, it remains unclear whether this is correlative or these alterations cause cancer. Importantly, Garcia et al. [93] demonstrated that increased expression of FACT subunits induces cancer in breast cell line, indicating that upregulation of FACT subunits can initiate breast cancer. However, this needs to be validated in in vivo or animal model. Further, these studies need to be expanded to other cancers or cancer types. Furthermore, it needs to be investigated if deletion and/or aforementioned mutations of FACT subunits can induce cancer.

Recently, Kaja et al. [20] demonstrated that increased expression/level of SPT16 enhances cellular growth and alters expression of the genes including the ones associated cancer. These results indicate that SPT16 amplification can induce cancer, consistent with aforementioned in vitro results of Garcia et al. [93]. However, it remains quite unclear how SPT16 is upregulated/amplified. Our recent studies implied impaired ubiquitylation and proteasomal degradation of SPT16 in its upregulation in cancer cells [20]. However, there can be additional mechanisms for SPT16 upregulation in cancer. Since our recent studies [20] indicated ubiquitylation and proteasomal degradation in Spt16 upregulation in cancer cells, the associated enzymes such as E2 conjugase, E3 ligase, and ubiquitin protease can emerge as important cancer biomarkers with targeted therapeutic potentials. However, the enzymes associated with the ubiquitylation and proteasomal degradation of SPT16 in humans are yet unknown. Future studies are warranted to identify these enzymes. In addition to be controlled by ubiquitylation and proteasomal degradation, SPT16 is regulated by other posttranslational modifications, which could be changed in cancer cells, leading to its altered level. Further, epigenetic, transcription, or post-transcription regulation of SPT16 could be changed in cancer, leading to its alteration. These possibilities remain to be investigated to deeply understand how SPT16 is regulated and what goes wrong in cancer. SSRP1 could be similarly regulated, alteration of which would change its level in cancer. Further, it is unknown how FACT subunits are amplified or deleted in various cancers. We are beginning to understand the FACT regulation and its involvement in breast and other cancers. However, future studies are needed to

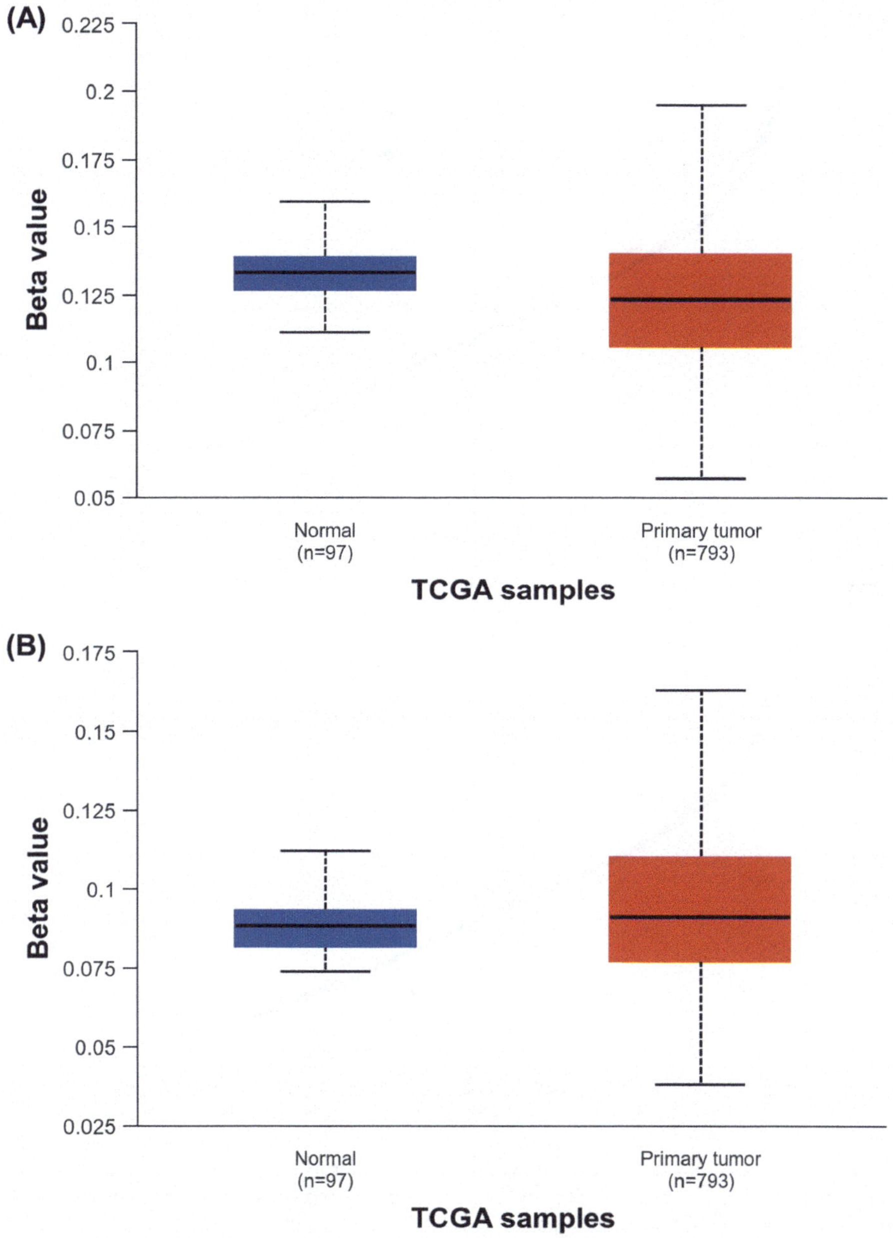


**Fig. 5.8** Analysis of promoter DNA methylation of the genes for SPT16 and SSRP1 in breast cancer patient samples. (**a**) Analysis of *SUPT16H* promoter DNA methylation in breast cancer. Beta value indicates the level of DNA methylation. It ranges from 0 (unmethylated) to 1 (fully methylated). Beta value of 0.7–0.5 represents hypermethylation, while hypomethylation is represented by the beta value between 0.3 and 0.25. (**b**) Analysis of *SSRP1* promoter DNA methylation in breast cancer

address above questions toward better understanding of FACT regulation with implications in cancer pathogenesis and targeted therapeutic intervention.

**Acknowledgments** The work in the Bhaumik laboratory was supported by the grants from National Institutes of Health (2R15GM088798-03), Simmons Cancer Institute (Team Science Grant), and SIU School of Medicine (Research Seed and Near-miss grants).

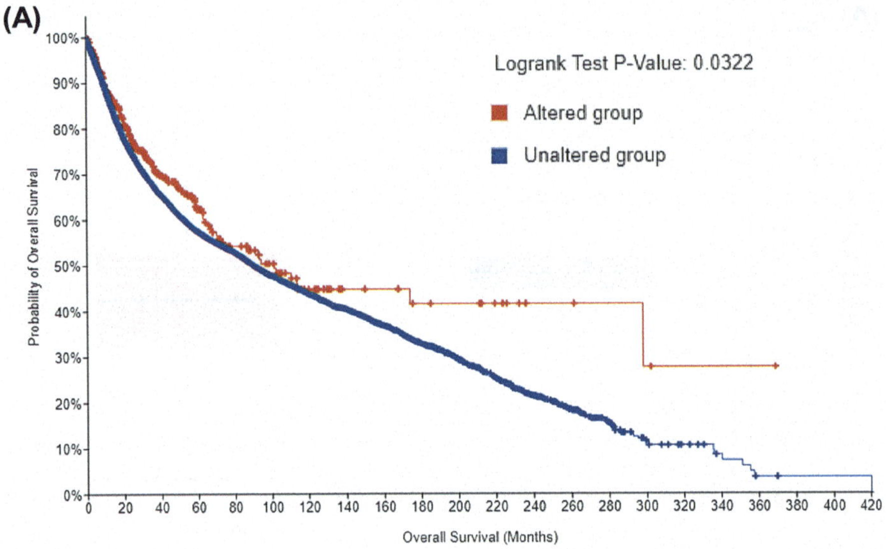


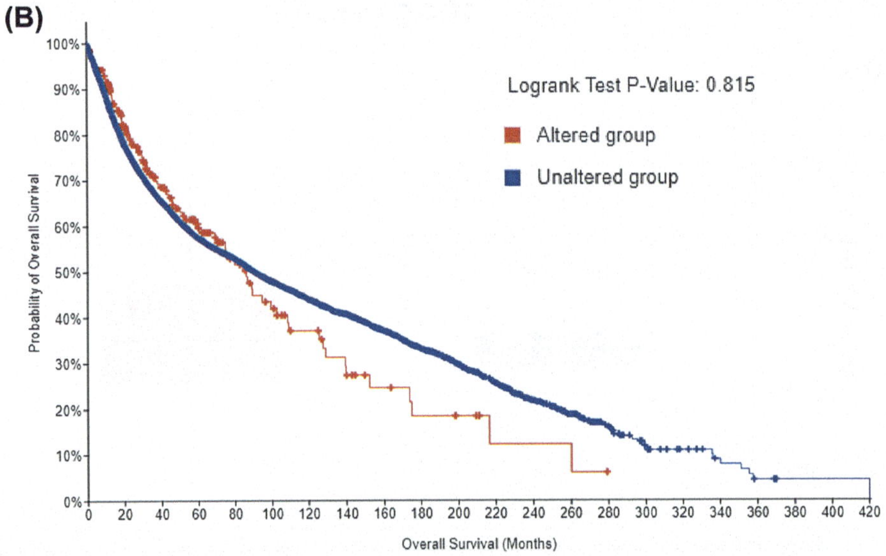


**Fig. 5.9** The overall survival probability with alterations of SPT16 and SSRP1 using a total of 70,655 samples/67,030 patients from 217 studies in cBioPortal. (**a**) Overall survival probability with alterations of SPT16. (**b**) Overall survival probability with alterations of SSRP1

## References


1. Luger K, Mader AW, Richmond RK, Sargent DF, Richmond TJ (1997) Crystal structure of the nucleosome core particle at 2.8 A resolution. Nature, 389: 251–260.
2. Kornberg RD, Lorch Y (2020) Primary role of the nucleosome. Mol. Cell., 79: 371–375.
3. Li B, Carey M, Workman JL (2007) The role of chromatin during transcription. Cell, 128: 707–719.
4. Kujirai T, Kurumizaka H (2020) Transcription through the nucleosome. Struct. Biol., 61: 42–49.
5. Bannister AJ, Kouzarides T (2011) Regulation of chromatin by histone modifications. Cell Res., 21: 381–395
6. Clapier CR, Iwasa J, Cairns BR, Peterson CL (2017) Mechanisms of action and regulation of ATP-dependent chromatin-remodelling complexes, Mol. Cell Biol.,18: 407–422

7. Gurard-Levin ZA, Quivy JP, Almouzni G (2014) Histone chaperones: assisting histone traffic and nucleosome dynamics. Biochem., 83: 487–517
8. Hammond CM Stromme CB, Huang H, Patel DJ, Groth A (2017) Histone chaperone networks shaping chromatin function. Mol. Cell. Biol., 18: 141–158
9. Martire S, Banaszynski LA (2020) The roles of histone variants in fine-tuning chromatin organization and function. Mol. Cell. Biol., 21: 522–541
10. Wu X, Zhang Y (2017) TET-mediated active DNA demethylation: mechanism, function and beyond. Nat. Rev. Genet., 18: 517–534
11. Greenberg MVC, Bourc'his D (2019) The diverse roles of DNA methylation in mammalian development and disease. Nat. Rev. Mol. Cell Biol., 20: 590–607
12. Grasser KD (2020) The FACT histone chaperone: Tuning gene transcription in the chromatin context to modulate plant growth and development. Front Plant Sci., 11:85.
13. Formosa T (2013) The role of FACT in making and breaking nucleosomes. Biochim Biophys Acta., 1819: 247–55.
14. Reinberg D, Sims 3rd R J (2006) de FACT Nucleosome Dynamics. J. Biol. Chem., 281: 23297–23301.
15. Formosa T, Winston F (2020) The role of FACT in managing chromatin: disruption, assembly, or repair? Nucleic Acids Res., 48: 11929–11941.
16. Wang P, Yang W, Zhao S, Nashun B (2021) Regulation of chromatin structure and function: insights into the histone chaperone FACT. Cell Cycle, 20: 465–479.
17. Jeronimo C, Robert F (2022) The histone chaperone FACT: a guardian of chromatin structure integrity. Transcription.; 13:16–38.
18. Gurova K., Chang HW, Valieva ME, Sandlesh P, Studitsky VM (2018) Structure and function of the histone chaperone FACT – Resolving FACTual issues. Biochim Biophys Acta Gene Regul Mech., 25: S1874–9399
19. Orphanides G, LeRoy G, Chang CH, Luse DS, Reinberg D (1998) FACT, a factor that facilitates transcript elongation through nucleosomes. Cell, 92: 105–116
20. Kaja A, Adhikari A, Karmakar S, Zhang W, Rothschild G, Basu U, Batra SK, Davie JK, Bhaumik SR (2021) Proteasomal regulation of mammalian Spt116 in controlling transcription. Mol. Cell. Biol., 41: e00452–20.
21. Koman IE, Commane M, Paszkiewicz G, Hoonjan B, Pal S, Safina A, Toshkov I, Purmal AA, Wang D, Liu S, Morrison C, Gudkov AV, Gurova KV (2012) Targeting FACT complex suppresses mammary tumorigenesis in Her2/neu transgenic mice. Cancer Prev Res (Phila), 5: 1025–35.
22. Fleyshman D, Prendergast L, Safina A, Paszkiewicz G, Commane M, Morgan K, Attwood K, Gurova K (2017) Level of FACT defines the transcriptional landscape and aggressive phenotype of breast cancer cells. Oncotarget, 8: 20525–20542.
23. Attwood K, Fleyshman D, Prendergast L, Paszkiewicz G, Omilian AR, Bshara W, Gurova K (2017) Prognostic value of histone chaperone FACT subunits expression in breast cancer. Breast Cancer (Dove Med Press), 9: 301–311.
24. Orphanides G, Wu WH, Lane WS et al (1999) The chromatin-specific transcription elongation factor FACT comprises human SPT16 and SSRP1 proteins. Nature, 400:284–288.
25. Belotserkovskaya R, Oh S, Bondarenko VA et al (2003) FACT facilitates transcription-dependent nucleosome alteration. Science, 301: 1090–1093.
26. Formosa T, Eriksson P, Wittmeyer J et al (2001) Spt16-Pob3 and the HMG protein Nhp6 combine to form the nucleosome-binding factor SPN. EMBO J., 20: 3506–3517.
27. Winkler DD, Luger K (2011) The histone chaperone FACT: structural insights and mechanisms for nucleosome reorganization. J Biol Chem., 286: 18369–18374
28. Hsieh FK, Kulaeva OI, Patel SS et al (2013) Histone chaperone FACT action during transcription through chromatin by RNA polymerase II. Proc Natl Acad Sci U S A, 110: 7654–7659.
29. Kemble DJ, McCullough LL, Whitby FG et al (2015) FACT disrupts nucleosome structure by binding H2A-H2B with conserved peptide motifs. Mol Cell, 60: 294–306
30. Hondele M, Stuwe T, Hassler M et al (2013) Structural basis of histone H2A-H2B recognition by the essential chaperone FACT. Nature, 499:111–114.
31. Valieva ME, Armeev GA, Kudryashova KS et al (2016) Large-scale ATP-independent nucleosome unfolding by a histone chaperone. Nat Struct Mol Biol., 23: 1111–1116.
32. Hoffmann C, Neumann H (2015) In Vivo mapping of FACT-histone interactions identifies a role of Pob3 C-terminus in H2A-H2B binding. ACS Chem Biol., 10: 2753–2763.
33. Tsunaka Y, Fujiwara Y, Oyama T et al (2016) Integrated molecular mechanism directing nucleosome reorganization by human FACT. Genes Dev., 30: 673–686.
34. Stuwe T, Hothorn M, Lejeune E et al (2008) The FACT Spt16 "peptidase" domain is a histone H3-H4 binding module. Proc Natl Acad Sci U S A., 105: 8884–8889.
35. Marciano G, Huang DT (2016) Structure of the human histone chaperone FACT Spt16 N-terminal domain. Acta Crystallogr F Struct Biol Commun., 72: 121–128.
36. Jamai A, Puglisi A, Strubin M (2009) Histone chaperone spt16 promotes redeposition of the original h3–h4 histones evicted by elongating RNA polymerase. Mol Cell, 35: 377–383.
37. Liu Y, Zhou K, Zhang N, Wei H, Tan YZ, Zhang Z, Carragher B, Potter CS., D'Arcy S, Luger K (2020).

FACT caught in the act of manipulating the nucleosome. Nature, 577, 426–431.
38. Mayanagi K, Saikusa K, Miyazaki N, Akashi S, Iwasaki K, Nishimura Y, Morikawa K, Tsunaka Y (2019). Structural visualization of key steps in nucleosome reorganization by human FACT. Sci Rep-uk 9: 10183.
39. Shirakata M, Huppi K, Usuda S, Okazaki K, Yoshida K, Sakano H (1991) HMG1-related DNA-binding protein isolated with V-(D)-J recombination signal probes, Mol Cell Biol., 11: 4528–4536.
40. Bruhn SL, Housman DE, Lippard SJ (1993) Isolation and characterization of cDNA clones encoding the Drosophila homolog of the HMG-box SSRP family that recognizes specific DNA structures, Nucleic Acids Res, 21: 1643–1646.
41. Bruhn SL, Pil PM, Essigmann JM, Housman DE, Lippard SJ (1992) Isolation and characterization of human cDNA clones encoding a high mobility group box protein that recognizes structural distortions to DNA caused by binding of the anticancer agent cisplatin, Proc Natl Acad Sci U S A, 89: 2307–2311.
42. Mason PB, Struhl K (2003) The FACT complex travels with elongating RNA polymerase II and is important for the fidelity of transcriptional initiation in vivo. Mol Cell Biol., 23: 8323–33.
43. Ransom M, Williams SK, Dechassa ML, Das C, Linger J, Adkins M, Liu C, Bartholomew B, Tyler JK (2009) FACT and the proteasome promote promoter chromatin disassembly and transcriptional initiation. J Biol Chem., 284: 23461–71.
44. Biswas D, Dutta-Biswas R, Stillman DJ (2007) Chd1 and yFACT act in opposition in regulating transcription. Mol Cell Biol., 27:6279–87.
45. Lolis AA, Londhe P, Beggs BC, Byrum SD, Tackett AJ, Davie JK (2013) Myogenin recruits the histone chaperone facilitates chromatin transcription (FACT) to promote nucleosome disassembly at muscle-specific genes. J Biol Chem., 288:7676–7687.
46. Tettey TT, Gao X, Shao W, Li H, Story BA, Chitsazan AD, Glaser RL, Goode ZH, Seidel CW, Conaway RC, Zeitlinger J, Blanchette M, Conaway JW (2019). A role for FACT in RNA polymerase II promoter-proximal pausing. Cell Rep, 27: 3770–3779.
47. Pathak R, Singh P, Ananthakrishnan S, Adamczyk S, Schimmel O, Govind CK (2018) Acetylation-dependent recruitment of the FACT complex and its role in regulating Pol II occupancy genome-wide in *Saccharomyces cerevisiae*. Genetics, 209: 743–756.
48. Fleming AB, Kao CF, Hillyer C, Pikaart M, Osley MA (2008) H2B ubiquitylation plays a role in nucleosome dynamics during transcription elongation. Mol Cell, 31: 57–66.
49. Biswas D, Yu Y, Prall M, Formosa T, Stillman DJ. (2005) The yeast FACT complex has a role in transcriptional initiation. Mol Cell Biol., 25: 5812–22.
50. Biswas D, Dutta-Biswas R, Mitra D, Shibata Y, Strahl BD, Formosa T, Stillman DJ (2006) Opposing roles for Set2 and yFACT in regulating TBP binding at promoters. EMBO J., 25: 4479–89.
51. Voth WP, Takahata S, Nishikawa JL, Metcalfe BM, Näär AM, Stillman DJ (2014) A role for FACT in repopulation of nucleosomes at inducible genes. PLoS One, 9: e84092.
52. Attwaters M (2023) Cryptic initiation drives transcriptional junk in ageing. Nat Rev Genet., 24: 345.
53. Sen P, Dang W, Donahue G, Dai J, Dorsey J, Cao X, Liu W, Cao K, Perry R, Lee JY, Wasko BM, Carr DT, He C, Robison B, Wagner J, Gregory BD, Kaeberlein M, Kennedy BK, Boeke JD, Berger SL (2015) H3K36 methylation promotes longevity by enhancing transcriptional fidelity. Genes Dev., 29: 1362–76.
54. McCauley BS, Sun L, Yu R, Lee M, Liu H, Leeman DS, Huang Y, Webb AE, Dang W (2021) Altered chromatin states drive cryptic transcription in aging mammalian stem cells. Nat Aging, 1: 684–697.
55. Sen P, Donahue G, Li C, Egervari G, Yang N, Lan Y, Robertson N, Shah PP, Kerkhoven E, Schultz DC, Adams PD, Berger SL (2023) Spurious intragenic transcription is a feature of mammalian cellular senescence and tissue aging. Nat Aging, 3: 402–417.
56. McCauley BS and Dang W (2022) Loosening chromatin and dysregulated transcription: A perspective on cryptic transcription during mammalian aging. Briefings in Functional Genomics, 21: 55–60.
57. Bhaumik SR, Malik S (2008) Diverse regulatory mechanisms of eukaryotic transcriptional activation by the proteasome complex. Crit Rev Biochem Mol Biol., 43: 419–33.
58. Frankland-Searby S, Bhaumik SR (2012) The 26S proteasome complex: an attractive target for cancer therapy. Biochim Biophys Acta, 1825: 64–76.
59. Bhaumik SR. (2011) Distinct regulatory mechanisms of eukaryotic transcriptional activation by SAGA and TFIID. Biochim Biophys Acta., 1809: 97–108.
60. Barman P, Chakraborty P, Bhaumik R, Bhaumik SR (2023) UPS writes a new saga of SAGA. Biochim Biophys Acta Gene Regul Mech., 1866: 194981.
61. Shukla A, Stanojevic N, Duan Z, Sen P, Bhaumik SR (2006) Ubp8p, a histone deubiquitinase whose association with SAGA is mediated by Sgf11p, differentially regulates lysine 4 methylation of histone H3 in vivo. Mol Cell Biol., 26: 3339–52.
62. Barman P, Kaja A, Chakraborty P, Guha S, Roy A, Ferdoush J, Bhaumik SR (2023) A novel ubiquitin-proteasome system regulation of Sgf73/ataxin-7 that maintains the integrity of the coactivator SAGA in orchestrating transcription. Genetics, 224: iyad071.
63. Fetian T, McShane BM, Horan NL, Shodja DN, True JD, Mosley AL, Arndt KM (2023) Paf1 complex subunit Rtf1 stimulates H2B ubiquitylation by interacting with the highly conserved N-terminal helix of Rad6. Proc Natl Acad Sci U S A., 120: e2220041120.
64. Chen F, Liu B, Guo L, Ge X, Feng W, Li DF, Zhou H, Long J (2021) Biochemical insights into Paf1 complex-induced stimulation of Rad6/Bre1-mediated H2B monoubiquitination. Proc Natl Acad Sci U S A., 118: e2025291118.

65. Kim J, Roeder RG (2009) Direct Bre1-Paf1 complex interactions and RING finger-independent Bre1-Rad6 interactions mediate histone H2B ubiquitylation in yeast. J Biol Chem., 284: 20582–92.
66. Van Oss SB, Shirra MK, Bataille AR, Wier AD, Yen K, Vinayachandran V, Byeon IL, Cucinotta CE, Héroux A, Jeon J, Kim J, VanDemark AP, Pugh BF, Arndt KM (2016) The histone modification domain of Paf1 complex subunit Rtf1 directly stimulates H2B ubiquitylation through an interaction with Rad6. Mol Cell, 64: 815–825.
67. Mueller JE, Canze M, Bryk M (2006) The requirements for COMPASS and Paf1 in transcriptional silencing and methylation of histone H3 in Saccharomyces cerevisiae. Genetics, 173: 557–67.
68. Krogan NJ, Dover J, Wood A, Schneider J, Heidt J, Boateng MA, Dean K, Ryan OW, Golshani A, Johnston M, Greenblatt JF, Shilatifard A (2003) The Paf1 complex is required for histone H3 methylation by COMPASS and Dot1p: linking transcriptional elongation to histone methylation. Mol Cell, 11: 721–9.
69. Bhaumik SR, Smith E, Shilatifard A (2007) Covalent modifications of histones during development and disease pathogenesis. Nat Struct Mol Biol., 14: 1008–16.
70. Shukla A, Chaurasia P, Bhaumik SR (2009) Histone methylation and ubiquitination with their cross-talk and roles in gene expression and stability. Cell Mol Life Sci., 66: 1419–33.
71. Malik S, Bhaumik SR (2010) Mixed lineage leukemia: histone H3 lysine 4 methyltransferases from yeast to human. FEBS J., 277: 1805–21.
72. Sen R, Bhaumik SR (2013) Transcriptional stimulatory and repressive functions of histone H2B ubiquitin ligase. Transcription, 4: 221–6.
73. Shukla A, Stanojevic N, Duan Z, Shadle T, Bhaumik SR (2006) Functional analysis of H2B-K123 ubiquitination in regulation of H3-K4 methylation and recruitment of RNA polymerase II at the coding sequences of several active genes in vivo. J. Biol. Chem., 281: 19045-19054.
74. Shukla A, Bhaumik SR (2007) H2B-K123 ubiquitination stimulates RNAPII elongation independent of H3-K4 methylation. Biochem Biophys Res Commun., 359: 214–20.
75. Hautbergue GM, Hung ML, Walsh MJ, Snijders AP, Chang CT, Jones R, Ponting CP, Dickman MJ, Wilson SA (2009) UIF, a New mRNA export adaptor that works together with REF/ALY, requires FACT for recruitment to mRNA. Curr Biol., 22: 1918–24.
76. Guha S, Barman P, Manawa A. Bhaumik SR (2022) Nuclear export of mRNAs with disease pathogenesis and therapeutic implications. Messenger RNA therapeutics, RNA Technologies 13, S. Jurga and J. Barciszewski (eds.), Springer Nature Switzerland AG 2022.
77. Herrera-Moyano E, Mergui X, García-Rubio ML, Barroso S, Aguilera A (2014) The yeast and human FACT chromatin-reorganizing complexes solve R-loop-mediated transcription-replication conflicts. Genes Dev., 28: 735–48
78. Sen R, Kaja A, Ferdoush J, Lahudkar S, Barman P, Bhaumik SR (2017) An mRNA capping enzyme targets FACT to the active gene to enhance the engagement of RNA polymerase II into transcriptional elongation. Mol Cell Biol., 37: e00029–17.
79. Deyter GM, Biggins S (2014) The FACT complex interacts with the E3 ubiquitin ligase Psh1 to prevent ectopic localization of CENP-A., Genes Dev., 16: 1815–26.
80. Han J, Li Q, McCullough L, Kettelkamp C, Formosa T, Zhang Z (2010) Ubiquitylation of FACT by the cullin-E3 ligase Rtt101 connects FACT to DNA replication. Genes Dev., 24: 1485–90.
81. Nakagawa T, Morohoshi A, Nagasawa Y, Nakagawa M, Hosogane M, Noda Y, Hosoi T, Nakayama K (2022) SPT16 ubiquitylation by DCAF14-CRL4 regulates FACT binding to histones. Cell Rep., 38: 110541.
82. Sen R, Ferdoush J, Kaja A, Bhaumik SR (2016) Fine-tuning of FACT by the ubiquitin proteasome system in regulation of transcriptional elongation. Mol Cell Biol., 36: 1691–703.
83. Shen J, Yang C, Zhang MS, Chin DW, Chan FF, Law CT, Wang G, Cheng CL, Chen M, Wan RT, Wu M, Kuang Z, Sharma R, Lee TKW, Ng IO, Wong CC, Wong CM (2022) Histone chaperone FACT complex coordinates with HIF to mediate an expeditious transcription program to adapt to poorly oxygenated cancers. Cell Rep., 38: 110304.
84. Tsunaka Y, Toga J, Yamaguchi H, Tate S-I, Hirose S, Morikawa K (2009) Phosphorylated intrinsically disordered region of FACT masks its nucleosomal DNA binding elements. J. Biol. Chem., 284: 24610–24621.
85. Murawska M, Greenstein RA, Schauer T, Olsen KCF, Ng H, Ladurner AG, Al-Sady B, Braun S (2021) The histone chaperone FACT facilitates heterochromatin spreading by regulating histone turnover and H3K9 methylation states., Cell Rep., 37: 109944.
86. Takahata S, Chida S, Ohnuma A, Ando M, Asanuma T, Murakami Y (2021) Two secured FACT recruitment mechanisms are essential for heterochromatin maintenance Cell Rep., 7: 109540.
87. Frost JM, Kim MY, Park GT, Hsieh PH, Nakamura M, Lin SJH, Yoo H, Choi J, Ikeda Y, Kinoshita T, Choi Y, Zilberman D, Fischer RL (2018) FACT complex is required for DNA demethylation at heterochromatin during reproduction in *Arabidopsis*. Proc Natl Acad Sci U S A., 115: E4720–E4729.
88. Yan X, Yang J, Xu J, Feng J, Li Q (2018) Histone chaperone Spt16p is required for heterochromatin mediated silencing in budding yeast. Protein Cell, 9: 652–658
89. Okada M, Okawa K, Isobe T, Fukagawa T (2009) CENP-H-containing complex facilitates centromere deposition of CENP-A in cooperation with FACT and CHD1. Mol Biol Cell., 20: 3986–95

90. Prendergast L, Müller S, Liu Y, Huang H, Dingli F, Loew D, Vassias I, Patel DJ, Sullivan KF, Almouzni G (2016) The CENP-T/-W complex is a binding partner of the histone chaperone FACT. Genes Dev., 30: 1313–26.
91. Lejeune E, Bortfeld M, White SA, Pidoux AL, Ekwall K, Allshire RC, Ladurner AG (2007) The chromatin-remodeling factor FACT contributes to centromeric heterochromatin independently of RNAi. Curr Biol.,17: 1219–24.
92. Garcia H, Miecznikowski JC, Safina A, Commane M, Ruusulehto A, Kilpinen S, Leach RW, Attwood K, Li Y, Degan S, Omilian AR, Guryanova O, Papantonopoulou O, Wang J, Buck M, Liu S, Morrison C, Gurova KV (2013) Facilitates chromatin transcription complex is an "accelerator" of tumor transformation and potential marker and target of aggressive cancers. Cell Rep., 4: 159–73.
93. Cerami E, Gao J, Dogrusoz U, Gross BE, Sumer SO, Aksoy BA, Jacobsen A, Byrne CJ, Heuer ML, Larsson E, Antipin Y, Reva B, Goldberg AP, Sander C, Schultz N (2012) The cBio Cancer Genomics Portal: an open platform for exploring multidimensional cancer genomics data. Cancer Discov., 2: 401–404.
94. Gao J, Aksoy BA, Dogrusoz U, Dresdner G, Gross B, Sumer SO, Sun Y, Jacobsen A, Sinha R, Larsson E, Cerami E, Sander C, Schultz N (2013) Integrative analysis of complex cancer genomics and clinical profiles using the cBioPortal. Sci Signal 6: pl1.
95. Cancer Genome Atlas Network (2012) Comprehensive molecular portraits of human breast tumours. Nature, 490: 61–70.
96. Davis CF, Ricketts CJ, Wang M, Yang L, Cherniack AD, Shen H, Buhay C, Kang H, Kim SC, Fahey CC, Hacker KE, Bhanot G, Gordenin DA, Chu A, Gunaratne PH, Biehl M, Seth S, Kaipparettu BA, Bristow CA, Donehower LA, Wallen EM, Smith AB, Tickoo SK, Tamboli P, Reuter V, Schmidt LS, Hsieh JJ, Choueiri TK, Hakimi AA, The Cancer Genome Atlas Network, Chin L, Meyerson M, Kucherlapati R, Park W-Y, Robertson AG, Laird PW, Henske EP, Kwiatkowski DJ, Park PJ, Morgan M, Shuch B, Muzny D, Wheeler DA, Linehan WM, Gibbs RA, Rathmell WK, Creighton CJ (2014) The somatic genomic landscape of chromophobe renal cell carcinoma. Cancer Cell, 26: 319–330.
97. Chandrashekar DS, Bashel B, Balasubramanya SAH, Creighton CJ, Ponce-Rodriguez I, Chakravarthi BVSK, Varambally S (2017) UALCAN: a portal for facilitating tumor subgroup gene expression and survival analyses. Neoplasia, 19: 649–658.
98. Chen F, Chandrashekar DS, Varambally S, Creighton CJ (2019) Pan-cancer molecular subtypes revealed by mass-spectrometry-based proteomic characterization of more than 500 human cancers. Nat Commun, 10: 5679.
99. Lolas IB, Himanen K, Grønlund JT, Lynggaard C, Houben A, Melzer M, van Lijsebettens M, Grasser KD (2010) The transcript elongation factor FACT affects *Arabidopsis* vegetative and reproductive development and genetically interacts with *HUB1/2*. Plant J, 61:686–697.
100. Ikeda Y, Kinoshita Y, Susaki D, Iwano M, Takayama S, Higashiyama T, Kakutani T, Kinoshita T (2011) HMG domain containing SSRP1 is required for DNA demethylation and genomic imprinting in *Arabidopsis*. Dev. Cell, 21: 589–596.

# Advances in Epigenetic Therapeutics for Breast Cancer

6

Michel-Gael F. Guefack and Sanchita Bhatnagar

**Abstract**

The epigenetic deregulations correlate with tumorigenesis, resistance to therapy, and metastasis of breast cancer cells. Given the predominance of aberrant epigenomic mechanisms, there is a growing emphasis on targeting epigenetic mechanisms for breast cancer therapeutic development. Selective inhibitors of epigenetic enzymes and the combined approach of epigenetic therapies with chemotherapies or hormone therapies in the treatment of breast cancer represent promising therapeutic strategies. In this chapter, we review the targeting of epigenetic mechanisms and highlight current epigenetic research in the development of breast cancer therapy.

**Keywords**

Azacitidine · Decitabine · Histone deacetylase inhibitors · Histone methyltransferase (HMT) inhibitors · Histone demethylase (HDM) inhibitors · Bromodomain and extra terminal domain (BET) family inhibitors

## 6.1 Introduction

Breast cancer remains one of the most frequently diagnosed cancers, with 313,510 estimated new cases and 42,780 deaths in the United States for the year 2024 [1]. Breast carcinogenesis is a multistep process caused by genetic and epigenetic abnormalities resulting in altered tumor suppressors and proto-oncogenes functions [2, 3]. With the increased awareness and availability of mammogram screenings, most breast cancers are diagnosed as palpable masses and localized to the breast tissue. In the United States, approximately 62% of breast cancers are confined to the breast at diagnosis, and only 6% of breast cancers are identified as metastatic [4]. The standard care for nonmetastatic breast cancer patients includes a combination of surgical resection of the tumor mass with systemic or localized postoperative therapies. The localized therapy mainly includes radiation treatment. The systemic treatments include chemotherapy (anthracycline, taxane, and the vinorelbine-capecitabine combination), endocrine therapies (antiestrogen hormone therapies), and treatment against the human epidermal growth factor receptor (HER) [5, 6]. Systemic therapies are the only treatment options available to metastatic breast cancer patients, which often are inefficient and have severe side effects.

The tumor tissue exhibits significant cellular heterogeneity owing to differential gene-specific transcriptional states and chromatin organization,

M.-G. F. Guefack · S. Bhatnagar (✉)
Department of Medical Microbiology and Immunology, University of California Davis School of Medicine, Davis, CA, USA
e-mail: sbbhatnagar@ucdavis.edu





S. Bhatnagar (ed.), *Epigenetic Mechanisms in Breast Cancer Therapy and Resistance*, Advances in Experimental Medicine and Biology 1465, https://doi.org/10.1007/978-3-031-66686-5_6

mainly attributed to aberrant epigenetic mechanisms [7]. The different epigenetic mechanisms in general transcription and their alteration in breast cancer have been described in the previous chapters. Given the predominance of aberrant epigenomic mechanisms, DNA methylation, and histone modifications in breast cancer, there is a growing emphasis on targeting epigenetic mechanisms for breast cancer therapeutics. The use of epigenetic therapies to alter resistance to chemotherapy also holds great promise in the clinical management of breast cancer [8]. Several epigenetic therapeutics in combination with endocrine therapy, chemotherapy, targeted therapies, radiotherapy, and immunotherapy are in clinical trials for the management of different breast cancer subtypes. These approved epigenetic anticancer agents, discussed below, target key players in epigenetic modification.

## 6.2 DNMT Inhibitors

One of the most widely studied epigenetic modifications is DNA methylation, which reversibly regulates the transcriptional state of a gene. DNA methylation is associated with worse overall patient survival through overexpression of oncogenes or downregulating tumor suppressors. Genome-wide hypomethylation is a characteristic feature of TNBC tumors [9]. For example, hypomethylation of the CpG island is linked to the TET1 overexpression and activation of oncogenic pathways, including PI3K, EGFR, and PDGF [10]. Notably, DNMT1 expression is upregulated in breast carcinoma patients compared to paired normal breast tissues [11]. The altered DNA methylation levels are associated with poor overall survival and breast cancer risk [12].

DNMT1 inhibitors are classified into nucleoside or non-nucleoside analogs [13]. The non-nucleoside analogs sequester DNMTs. On the other hand, the nucleosides induce DNA hypomethylation by targeting DNMTs for proteasomal degradation and act as cytosine mimics. The small molecules azacitidine and decitabine are approved for the treatment of hematological malignancies, but their efficiency in breast cancer as a single agent is limited. A phase I study tested a combination of azacitidine and valproic acid, which stabilized the disease in a small cohort of patients [14]. In contrast, a phase II clinical trial combining the histone deacetylase inhibitors (HDACi) entinostat and azacitidine did not benefit the TNBC patients [15].

The DNMT inhibitors decitabine and azacitidine can also activate antitumor immune effector cells, promote antigen processing machinery, and induce the expression of cytokines and tumor antigens [16]. Notably, the DNMT inhibitors also affect PD1 signaling via modulation of PD-L1 expression and PD-1 blockade [17, 18]. An ongoing phase II trial investigates the efficacy of neoadjuvant pembrolizumab (anti-PD-1) plus decitabine in TNBC and HER-/HR + patients (NCT02957968).

A second-generation DNMT inhibitor, guadecitabine (SGI-110), is a derivative of decitabine with improved plasma half-life and efficiency [8, 11, 19]. The combination of HDACi and guadecitabine reprograms aggressive TNBC cells to a less aggressive phenotype by reversing EMT [19].

In addition to small molecules, natural products bind DNMT1 with antitumor effects in TNBC cells. Antroquinonol D (3-dimethoxy antroquinonol) is a compound derived from the mushroom *Antrodia camphorata*, which competes with the cofactor S-adenosyl methionine (SAM) and inhibits the catalytic activity of DNMT1 [20]. Kazinol Q, a compound derived from Formosan plants, inhibits DNMT1 by competing with the cytosine binding [21]. Isofistularin-3 is isolated from the marine sponge *Aplysina aerophoba* and inhibits the enzymatic activity of DNMT [22].

## 6.3 Histone Deacetylase Inhibitors Inhibitors (HDACi)

HDACs are a class of enzymes that catalyze the removal of the acetyl groups covalently linked to the lysine residues of histone proteins. The overexpression of HDAC in breast cancer cells induces transcriptional silencing of the ERα hormone receptor [23]. Histone deacetylase inhibitors are approved and licensed by the Food and Drug Administration (FDA) for cancer therapy. HDACi can be divided into five classes based on their chemical structures [24, 25]. Class I contains hydroxamic acids (suberoylanilide hydroxamic acid and trichostatin A). Several Class I HDACi are approved for clinical use, including vorinostat for cutaneous T-cell lymphoma, belinostat for peripheral T-cell lymphoma, and panobinostat for myeloma. The Class II HDACi are short-chain fatty acids, such as valproic acid, butyric acid, and phenylbutyric acid. Class III are benzamides that include entinostat, tacedinaline, and mocetinostat. The cyclic tetrapeptides constitute the Class IV HDACi and include romidepsin, which is approved for cutaneous T-cell lymphoma. The Class V contains sirtuin inhibitors, such as nicotinamide.

Mechanistically, HDACi induce programmed cell death in cancer cells. For example, HDACi upregulates the pro-apoptotic gene BMF but downregulates the BCL2A1 pro-survival gene in TNBC cells [26]. Trichostatin A induces cell cycle arrest at the G1 and G2 phases and suppresses the TNBC cancer cell proliferation in the culture [27, 28]. Valproic acid inhibits the cell cycle at the G1-S and G2-M phases via the cyclin-dependent kinase inhibitor p21 [29]. The antiproliferative activity of valproic acid is also linked to the hyperacetylation of HSP70 [30, 31]. The bortezomib and panobinostat restore ER expression in vitro by displacing DNMT1, HDAC1, and SUV39H1 [32]. While TNBC cancer treatments are limited due to the lack of ER hormone receptors, re-expressing hormone receptors to sensitize TNBC cells has been explored. The entinostat treatment of previously ER-negative breast cancer cells induced ER receptor re-expression in vitro [33].

## 6.4 Histone Methyltransferase (HMT) Inhibitors

HMTs play crucial roles in epigenetic regulation by catalyzing the addition of methyl groups to lysine and arginine residues on histone tails and other protein targets. At the level of active genes, specific methylation patterns on histone H3 and H4 are associated with either gene activation or repression. For instance, mono-methylation of residues such as H3K4, H3K9, H3K27, H3K36, and H4K20 is generally linked to gene activation. Meanwhile, di-methylation of H3K79 and trimethylation of H3K36 are also associated with active gene expression [34, 35]. One of the most studied mechanisms of repressive histone methylation involves trimethylation of H3K27, which is mediated by the polycomb repressive complex 2 (PRC2). In mammals, the core PRC2 complex comprises several subunits, including EED, SUZ12, NURF55, Rbap46/48, and one of the catalytic subunits, either enhancer of zeste homolog 1 (EZH1) or enhancer of zeste homolog 2 (EZH2) [36, 37].

The elevated expression of HMTs in breast cancer patients is linked to acquired resistance to treatment and progression, indicating their potential as predictive biomarkers and therapeutic targets. Pharmacological targeting of methyltransferases, EZH2, G9a, disruptor of telomeric silencing 1-like protein (DOT1L), and protein arginine methyltransferases (PRMT) 1 and 5 are currently being evaluated as treatment options for breast cancer patients. Several selective EZH2 inhibitors targeting H3K27 di- and trimethylation include CPI-0209, CPI-1205, tazemetostat (EPZ-6438), GSK2816126, PF-06821497, and DS-3201 [37, 38]. While tazemetostat is approved for epithelioid sarcoma [39], the clinical trial of GSK2816126 showed modest anticancer activity [40]. Selective inhibitors of DOT1L include EPZ004777, EPZ-5676 [pinometostat], and SGC0496, which showed limited success in clinical trials for MLL [41]. The PRMT5 inhibitors, GSK3326595 (EPZ015938), JNJ-6461978, and PF-06939999, are currently being evaluated for hematological malignancies and solid tumors [41].

Sinefungin A, a natural product isolated from *Streptomyces* spp., was identified as the SAM-competitive inhibitor of KMTs [42]. Chaetocin inhibits the *Drosophila melanogaster* Suv39 family, including Suv39h1, and shows selectivity against certain KMTs. Chaetocin has demonstrated anti-myeloma activity and can inhibit Suv39h1 in AML cells with hypermethylated tumor-suppressor genes [43, 44]. Furthermore,

JMF034, a gliotoxin analog isolated from Penicillium sp., is a potent inhibitor of G9a and Suv39h1 [45].

## 6.5 Histone Demethylase (HDM) Inhibitors

HDMs regulate gene expression by removing methyl groups from histone proteins, influencing chromatin structure and gene transcription [46]. Dysregulation of HDM has been implicated in various cancers, including breast cancer. Thus, inhibition of HDMs reverses aberrant gene expression patterns associated with cancer progression. Several histone demethylase inhibitors have been studied in the context of breast cancer and represent powerful candidates for involvement in clinical therapy. GSK-J1 is a potent and selective inhibitor of histone demethylase JMJD3 [47]. GSK-J1 inhibits the growth of breast cancer cells in vitro and in vivo by inducing differentiation and inhibiting cell proliferation [48].

IOX1 is a broad-spectrum inhibitor of histone demethylases, including members of the Jumonji-C domain-containing (JmjC) family. IOX1 induces apoptosis to reduce TNBC cell invasion and metastasis [49]. CPI-455 is a selective inhibitor of lysine-specific demethylase 1 (LSD1) [50]. LSD1 is overexpressed in various cancers, including breast cancer, and its inhibition induces differentiation and inhibits proliferation in breast cancer cells [51]. SP2509 is another selective LSD1 inhibitor that inhibits the growth of breast cancer cells [50]. Interestingly, the co-administration of anti-PD1 and LSD1 inhibitors in TNBC mouse models showed an antitumor effect [52].

## 6.6 Bromodomain and Extra Terminal Domain (BET) Family Inhibitors

BET family proteins, including BRD2, BRD3, BRD4, and BRDT, are epigenetic readers for acetylated lysine residues on histones [53]. BET proteins promote oncogenic transcriptional programs in breast cancer, and several BET inhibitors have been developed for their therapeutic implications [54, 55].

BET151, also known as GSK1210151A, is a potent and selective inhibitor of BET proteins, particularly BRD2, BRD3, and BRD4. BET151 suppresses the expression of oncogenes such as MYC and cyclin D1 [54] and induces cell cycle arrest and apoptosis in breast cancer cells [53]. Preclinical studies indicated the efficacy of BET151 across different molecular subtypes of breast cancer, including hormone receptor positive, HER2 positive, and TNBC [56]. In addition, BET151 suppresses the invasive and migratory potential of breast cancer cells by targeting BRD4-mediated transcriptional programs involved in epithelial-mesenchymal transition (EMT) and metastasis [57]. BET151 also synergizes the effects of hormonal therapy, chemotherapy, and targeted therapies, and the combinatorial approach is expected to overcome resistance mechanisms in the breast cancer [58].

JQ1 is another well-studied BET inhibitor that competes with acetyl-lysine for binding to the bromodomain of BET proteins. Its role in breast cancer research has garnered significant interest due to its ability to disrupt oncogenic transcriptional programs driven by BET proteins. In breast cancer, JQ1 downregulates the expression of key oncogenes, such as MYC, which promote tumor growth, proliferation, and survival [59]. JQ1 inhibits tumor cell growth and promotes cell death in breast cancer cells by modulating genes involved in cell cycle regulation, apoptosis, and DNA damage response pathways [60, 61]. Preclinical studies have demonstrated the JQ1 efficacy across different molecular subtypes of breast cancer [60].

OTX015, also known as MK-8628, is another potent and orally active BET inhibitor. It functions by competitively binding to the bromodomain of BET proteins, disrupting their interaction with acetylated lysine residues on histones and consequently inhibiting oncogenic transcriptional programs. While OTX015 is tested for hematological malignancies, its potential role in breast cancer is also being evaluated [62]. OTX015 inhibits the expression of key oncogenes such as MYC, cyclin D1, and BCL2, which promote tumor proliferation, survival, and resistance to apoptosis [63]. Like other BET inhibitors, OTX015 is also effective across different molecular subtypes of breast cancer, including hormone receptor positive, HER2 positive, and TNBC [64].

## 6.7 Promising Epigenetic Therapy Combinations

Targeting epigenetic modifications in breast cancer is a promising therapeutic avenue and could help in overcoming resistance to conventional therapies. Combining epigenetic therapies with conventional or other targeted therapies can accentuate efficacy. Some promising epigenetic therapy combinations in breast cancer include HDACi with hormone therapy. HDACi, such as vorinostat or romidepsin combined with hormone therapy (e.g., tamoxifen or aromatase inhibitors), have shown synergistic effects in hormone receptor-positive breast cancer [65]. HDACi can reverse hormone therapy resistance by reactivating silenced estrogen receptor genes and, thus, sensitizing the cancer cells to endocrine therapies [66]. DNMT inhibitors, such as decitabine or azacitidine combined with chemotherapy drugs (e.g., doxorubicin or paclitaxel), have demonstrated efficacy in TNBC. DNMT inhibitors can sensitize TNBC cells to chemotherapy by reactivating tumor-suppressor genes silenced through DNA methylation. Combining HDACi with immune checkpoint inhibitors (e.g., pembrolizumab) has shown promise in enhancing the immune response against breast cancer cells. HDACi sensitizes cancer cells to immune-mediated killing by upregulating the expression of immune-related genes and increasing tumor immunogenicity [67].

PARP inhibitors such as olaparib and veliparib combined with HDACi synergize in breast cancer, particularly in BRCA-mutated tumors [68]. HDACi can enhance DNA damage induced by PARP inhibitors, leading to synthetic lethality in BRCA-deficient cancer cells [69].

BET inhibitors, in combination with cyclin-dependent kinase 4/6 (CDK4/6), such as palbociclib, synergistically affect hormone receptor-positive breast cancer [70]. BET inhibitors can downregulate MYC expression, enhancing the efficacy of CDK4/6 inhibitors to block cell cycle progression [54]. Combining HDACi with phosphoinositide 3-kinase (PI3K) inhibitors (e.g., alpelisib) has also shown promise in overcoming resistance to PI3K pathway-targeted therapies in HER2-positive breast cancer [71]. HDACi modulate PI3K signaling and restore sensitivity to PI3K inhibitors [72]. These combinatorial approaches represent a promising avenue for improving treatment outcomes in breast cancer patients by targeting epigenetic alterations in conjunction with standard therapies or other targeted agents.

## 6.8 Personalized Medicine in Breast Cancer

Precision medicine requires characterizing distinct genomic and transcriptomic profiles of breast tumors and using that information to formulate treatment regimens unique to a patient. Genetic and epigenetic data, including structural variations, gene expression profiles, DNA methylation profiles, histone modifications, and 3D genome structures, are crucial to personalized medicine. To this end, various multi-omics approaches, such as Oncotype DX, Breast Cancer Index, EndoPredict, MammaPrint, NanoString, and Prosigna Breast Cancer Prognostic Gene Signature Assay, are used [73, 74]. Chromatin accessibility, which influences DNA binding to transcription factors and regulatory elements, plays a crucial role in dictating the transcriptional states of oncogenes and tumor suppressors. ATAC-seq and ChIP-seq are used to interrogate the chromatin states genome-wide [75].

Furthermore, disrupting the genome's 3D structure can lead to oncogenic activation by altering interactions between enhancers, promoter regions, and distant regions that control oncogene expression. Genome-wide HiC mapping of interactions at a single-base resolution provides crucial insights into tumor biology. Integrating these epigenomic data with other sequencing techniques, such as whole genome sequencing and RNA-seq, identifies patient-specific genetic and epigenetic alterations and is central to precise and personalized treatment strategies. Integrating multi-omics approaches to identify diagnostic biomarkers represents a significant advancement in personalized medicine, particularly in the cancer [76, 77].

Artificial intelligence (AI) based on machine learning and deep learning has revolutionized precision medicine by providing powerful tools for analyzing and interpreting complex data. AI has automated the interpretation of medical images, such as histological sections of tumor tissues or X-ray images, enabling precise quantification of tumor characteristics. Deep neural networks are particularly effective in processing digital images to differentiate cancer cells from normal cells and distinguish benign from malignant tissue [78].

## 6.9 Adverse Effects of Epigenetic Therapy

Most established epigenetic drugs pose the general problem of target specificity, which could lead to a general transcriptional effect [79]. Off-target effects lead to unintended changes in gene expression and potentially harm body tissues and organs [80]. As such, epigenetic drugs can cause gastrointestinal problems, liver toxicity, or bone marrow suppression. The administration doses required to drive the specific anticancer mechanisms must also be considered in a combined therapy regimen to limit adverse effects.

Like conventional treatments, drug resistance can develop in cancer cells against epigenetic therapies, making the treatment less effective [81]. There is a risk of reactivating the silenced genes, leading to tumor progression or recurrence. Epigenetic reprogramming may also have long-term consequences or unintended effects on cell function [82]. Thus, the potential risks and side effects of epigenetic therapies should be carefully monitored and managed.

## 6.10 Conclusion

The development of drugs targeting epigenetic modulators represents a significant advance in cancer therapy. Combined strategies, which could include epigenetic therapies in tandem with other modalities such as chemotherapy or immunotherapy, could further benefit breast cancer patients. Understanding the epigenetic landscape of cancer is essential to guiding the development of personalized treatments and improving therapeutic outcomes. By integrating information on the cancer epigenome, precision medicine opens new avenues for more accurate diagnosis and more effective therapeutic strategies.

## References


1. Siegel, R.L., A.N. Giaquinto, and A. Jemal, *Cancer statistics, 2024*. CA Cancer J Clin, 2024. **74**(1): p. 12–49.
2. Lee, R.S., et al., *Emerging Role of Epigenetic Modifiers in Breast Cancer Pathogenesis and Therapeutic Response*. Cancers (Basel), 2023. **15**(15).
3. Collins, A. and I. Politopoulos, *The genetics of breast cancer: risk factors for disease*. Appl Clin Genet, 2011. **4**: p. 11–9.
4. Marks, C.E., et al., *Metastatic breast cancer: Who benefits from surgery?* Am J Surg, 2022. **223**(1): p. 81–93.
5. Rubovszky, G., et al., *Systemic Treatment of Breast Cancer. 1st Central-Eastern European Professional Consensus Statement on Breast Cancer*. Pathol Oncol Res, 2022. **28**: p. 1610383.
6. Agostinetto, E., J. Gligorov, and M. Piccart, *Systemic therapy for early-stage breast cancer: learning from the past to build the future*. Nat Rev Clin Oncol, 2022. **19**(12): p. 763–774.
7. Carter, B. and K. Zhao, *The epigenetic basis of cellular heterogeneity*. Nat Rev Genet, 2021. **22**(4): p. 235–250.
8. Strauss, J. and W.D. Figg, *Using Epigenetic Therapy to Overcome Chemotherapy Resistance*. Anticancer Res, 2016. **36**(1): p. 1–4.
9. Huang, T.H., M.R. Perry, and D.E. Laux, *Methylation profiling of CpG islands in human breast cancer cells*. Hum Mol Genet, 1999. **8**(3): p. 459–70.
10. Good, C.R., et al., *TET1-Mediated Hypomethylation Activates Oncogenic Signaling in Triple-Negative Breast Cancer*. Cancer Res, 2018. **78**(15): p. 4126–4137.
11. Wong, K.K., *DNMT1: A key drug target in triple-negative breast cancer*. Semin Cancer Biol, 2021. **72**: p. 198–213.

12. Salta, S., et al., *A DNA Methylation-Based Test for Breast Cancer Detection in Circulating Cell-Free DNA*. J Clin Med, 2018. **7**(11).
13. Mehdipour, P., R. Chen, and D.D. De Carvalho, *The next generation of DNMT inhibitors*. Nat Cancer, 2021. **2**(10): p. 1000–1001.
14. Braiteh, F., et al., *Phase I study of epigenetic modulation with 5-azacytidine and valproic acid in patients with advanced cancers*. Clin Cancer Res, 2008. **14**(19): p. 6296–301.
15. Connolly, R.M., et al., *Combination Epigenetic Therapy in Advanced Breast Cancer with 5-Azacitidine and Entinostat: A Phase II National Cancer Institute/Stand Up to Cancer Study*. Clin Cancer Res, 2017. **23**(11): p. 2691–2701.
16. Dan, H., et al., *DNA Methyltransferase Inhibitors: Catalysts For Antitumour Immune Responses*. Onco Targets Ther, 2019. **12**: p. 10903–10916.
17. Dai, M., et al., *New insights into epigenetic regulation of resistance to PD-1/PD-L1 blockade cancer immunotherapy: mechanisms and therapeutic opportunities*. Exp Hematol Oncol, 2022. **11**(1): p. 101.
18. Ma, X., et al., *Epigenetic modifications: Critical participants of the PD-L1 regulatory mechanism in solid tumors (Review)*. Int J Oncol, 2022. **61**(5).
19. Su, Y., et al., *Epigenetic reprogramming of epithelial mesenchymal transition in triple negative breast cancer cells with DNA methyltransferase and histone deacetylase inhibitors*. J Exp Clin Cancer Res, 2018. **37**(1): p. 314.
20. Wang, S.C., et al., *Antroquinonol D, isolated from Antrodia camphorata, with DNA demethylation and anticancer potential*. J Agric Food Chem, 2014. **62**(24): p. 5625–35.
21. Weng, J.R., et al., *Identification of kazinol Q, a natural product from Formosan plants, as an inhibitor of DNA methyltransferase*. Phytother Res, 2014. **28**(1): p. 49–54.
22. Florean, C., et al., *Discovery and characterization of Isofistularin-3, a marine brominated alkaloid, as a new DNA demethylating agent inducing cell cycle arrest and sensitization to TRAIL in cancer cells*. Oncotarget, 2016. **7**(17): p. 24027–49.
23. Linares, A., et al., *Increased expression of the HDAC9 gene is associated with antiestrogen resistance of breast cancers*. Mol Oncol, 2019. **13**(7): p. 1534–1547.
24. Yang, F., et al., *Next-generation of selective histone deacetylase inhibitors*. RSC Adv, 2019. **9**(34): p. 19571–19583.
25. Kim, H.J. and S.C. Bae, *Histone deacetylase inhibitors: molecular mechanisms of action and clinical trials as anti-cancer drugs*. Am J Transl Res, 2011. **3**(2): p. 166–79.
26. Bolden, J.E., et al., *HDAC inhibitors induce tumor-cell-selective pro-apoptotic transcriptional responses*. Cell Death Dis, 2013. **4**(2): p. e519.
27. Yang, X., et al., *Histone deacetylase inhibitor suberoyl bis-hydroxamic acid suppresses cell proliferation and induces apoptosis in breast cancer cells*. Mol Med Rep, 2015. **11**(4): p. 2908–12.
28. Wang, S.C., et al., *Trichostatin A induces bladder cancer cell death via intrinsic apoptosis at the early phase and Sp1-survivin downregulation at the late phase of treatment*. Oncol Rep, 2017. **38**(3): p. 1587–1596.
29. Giordano, F., et al., *Valproic acid inhibits cell growth in both MCF-7 and MDA-MB231 cells by triggering different responses in a cell type-specific manner*. J Transl Med, 2023. **21**(1): p. 165.
30. Ma, X.J., et al., *The role and possible molecular mechanism of valproic acid in the growth of MCF-7 breast cancer cells*. Croat Med J, 2017. **58**(5): p. 349–357.
31. Mawatari, T., et al., *Valproic acid inhibits proliferation of HER2-expressing breast cancer cells by inducing cell cycle arrest and apoptosis through Hsp70 acetylation*. Int J Oncol, 2015. **47**(6): p. 2073–81.
32. Al-Kabariti, A.Y. and M.A. Abbas, *Progress in the Understanding of Estrogen Receptor Alpha Signaling in Triple-Negative Breast Cancer: Reactivation of Silenced ER-alpha and Signaling through ER-alpha36*. Mol Cancer Res, 2023. **21**(11): p. 1123–1138.
33. Dewi, C., et al., *Signaling Pathways and Natural Compounds in Triple-Negative Breast Cancer Cell Line*. Molecules, 2022. **27**(12).
34. Onder, T.T., et al., *Chromatin-modifying enzymes as modulators of reprogramming*. Nature, 2012. **483**(7391): p. 598–602.
35. Zhang, T., S. Cooper, and N. Brockdorff, *The interplay of histone modifications – writers that read*. EMBO Rep, 2015. **16**(11): p. 1467–81.
36. Pal, B., et al., *Global changes in the mammary epigenome are induced by hormonal cues and coordinated by Ezh2*. Cell Rep, 2013. **3**(2): p. 411–26.
37. Katoh, M., *Mutation spectra of histone methyltransferases with canonical SET domains and EZH2-targeted therapy*. Epigenomics, 2016. **8**(2): p. 285–305.
38. Kim, K.H. and C.W. Roberts, *Targeting EZH2 in cancer*. Nat Med, 2016. **22**(2): p. 128–34.
39. Simeone, N., et al., *Tazemetostat for advanced epithelioid sarcoma: current status and future perspectives*. Future Oncol, 2021. **17**(10): p. 1253–1263.
40. McCabe, M.T., et al., *EZH2 inhibition as a therapeutic strategy for lymphoma with EZH2-activating mutations*. Nature, 2012. **492**(7427): p. 108–12.
41. Rugo, H.S., et al., *The Promise for Histone Methyltransferase Inhibitors for Epigenetic Therapy in Clinical Oncology: A Narrative Review*. Adv Ther, 2020. **37**(7): p. 3059–3082.
42. Wang, Z. and D.J. Patel, *Small molecule epigenetic inhibitors targeted to histone lysine methyltransferases and demethylases*. Q Rev Biophys, 2013. **46**(4): p. 349–73.
43. Tian, X., et al., *Histone lysine-specific methyltransferases and demethylases in carcinogenesis: new targets for cancer therapy and prevention*. Curr Cancer Drug Targets, 2013. **13**(5): p. 558–79.

44. Sak, A., et al., *Chaetocin induced chromatin condensation: effect on DNA repair signaling and survival.* Int J Radiat Biol, 2021. **97**(4): p. 494–506.
45. Goher, S.S., et al., *2,5-Diketopiperazines (DKPs): Promising Scaffolds for Anticancer Agents.* Curr Pharm Des, 2024. **30**(8): p. 597–623.
46. Kang, M.K., et al., *Epigenetic gene regulation by histone demethylases: emerging role in oncogenesis and inflammation.* Oral Dis, 2017. **23**(6): p. 709–720.
47. Dalpatraj, N., A. Naik, and N. Thakur, *GSK-J4: An H3K27 histone demethylase inhibitor, as a potential anti-cancer agent.* Int J Cancer, 2023. **153**(6): p. 1130–1138.
48. Abu-Hanna, J., et al., *Therapeutic potential of inhibiting histone 3 lysine 27 demethylases: a review of the literature.* Clin Epigenetics, 2022. **14**(1): p. 98.
49. Liu, J., et al., *Co-delivery of IOX1 and doxorubicin for antibody-independent cancer chemo-immunotherapy.* Nat Commun, 2021. **12**(1): p. 2425.
50. Das, N.D., H. Niwa, and T. Umehara, *Chemical Inhibitors Targeting the Histone Lysine Demethylase Families with Potential for Drug Discovery.* Epigenomes, 2023. **7**(1).
51. Benedetti, R., et al., *Inhibition of Histone Demethylases LSD1 and UTX Regulates ERalpha Signaling in Breast Cancer.* Cancers (Basel), 2019. **11**(12).
52. Nanamori, H. and Y. Sawada, *Epigenetic Modification of PD-1/PD-L1-Mediated Cancer Immunotherapy against Melanoma.* Int J Mol Sci, 2022. **23**(3).
53. Taniguchi, Y., *The Bromodomain and Extra-Terminal Domain (BET) Family: Functional Anatomy of BET Paralogous Proteins.* Int J Mol Sci, 2016. **17**(11).
54. Shorstova, T., W.D. Foulkes, and M. Witcher, *Achieving clinical success with BET inhibitors as anti-cancer agents.* Br J Cancer, 2021. **124**(9): p. 1478–1490.
55. Doroshow, D.B., J.P. Eder, and P.M. LoRusso, *BET inhibitors: a novel epigenetic approach.* Ann Oncol, 2017. **28**(8): p. 1776–1787.
56. Andrikopoulou, A., et al., *The emerging role of BET inhibitors in breast cancer.* Breast, 2020. **53**: p. 152–163.
57. Lai, J., et al., *Anticancer Effects of I-BET151, an Inhibitor of Bromodomain and Extra-Terminal Domain Proteins.* Front Oncol, 2021. **11**: p. 716830.
58. Riobo-Del Galdo, N.A., A. Lara Montero, and E.V. Wertheimer, *Role of Hedgehog Signaling in Breast Cancer: Pathogenesis and Therapeutics.* Cells, 2019. **8**(4).
59. Jiang, G., et al., *General mechanism of JQ1 in inhibiting various types of cancer.* Mol Med Rep, 2020. **21**(3): p. 1021–1034.
60. Shi, X., et al., *JQ1: a novel potential therapeutic target.* Pharmazie, 2018. **73**(9): p. 491–493.
61. da Motta, L.L., et al., *The BET inhibitor JQ1 selectively impairs tumour response to hypoxia and downregulates CA9 and angiogenesis in triple negative breast cancer.* Oncogene, 2017. **36**(1): p. 122–132.
62. Berthon, C., et al., *Bromodomain inhibitor OTX015 in patients with acute leukaemia: a dose-escalation, phase 1 study.* Lancet Haematol, 2016. **3**(4): p. e186–95.
63. Megiorni, F., et al., *OTX015 Epi-Drug Exerts Antitumor Effects in Ovarian Cancer Cells by Blocking GNL3-Mediated Radioresistance Mechanisms: Cellular, Molecular and Computational Evidence.* Cancers (Basel), 2021. **13**(7).
64. Ocana, A., C. Nieto-Jimenez, and A. Pandiella, *BET inhibitors as novel therapeutic agents in breast cancer.* Oncotarget, 2017. **8**(41): p. 71285–71291.
65. Jenke, R., et al., *Anticancer Therapy with HDAC Inhibitors: Mechanism-Based Combination Strategies and Future Perspectives.* Cancers (Basel), 2021. **13**(4).
66. Dimitrakopoulos, F.I., A. Kottorou, and A. Tzezou, *Endocrine resistance and epigenetic reprogramming in estrogen receptor positive breast cancer.* Cancer Lett, 2021. **517**: p. 55–65.
67. Hogg, S.J., et al., *Targeting the epigenetic regulation of antitumour immunity.* Nat Rev Drug Discov, 2020. **19**(11): p. 776–800.
68. Zhu, Q., et al., *Novel dual inhibitors of PARP and HDAC induce intratumoral STING-mediated antitumor immunity in triple-negative breast cancer.* Cell Death Dis, 2024. **15**(1): p. 10.
69. Marijon, H., et al., *Co-targeting poly(ADP-ribose) polymerase (PARP) and histone deacetylase (HDAC) in triple-negative breast cancer: Higher synergism in BRCA mutated cells.* Biomed Pharmacother, 2018. **99**: p. 543–551.
70. Manzotti, G., A. Ciarrocchi, and V. Sancisi, *Inhibition of BET Proteins and Histone Deacetylase (HDACs): Crossing Roads in Cancer Therapy.* Cancers (Basel), 2019. **11**(3).
71. Wang, Y. and A. Minden, *Current Molecular Combination Therapies Used for the Treatment of Breast Cancer.* Int J Mol Sci, 2022. **23**(19).
72. Rodrigues, D.A., P.S.M. Pinheiro, and C.A.M. Fraga, *Multitarget Inhibition of Histone Deacetylase (HDAC) and Phosphatidylinositol-3-kinase (PI3K): Current and Future Prospects.* ChemMedChem, 2021. **16**(3): p. 448–457.
73. Vieira, A.F. and F. Schmitt, *An Update on Breast Cancer Multigene Prognostic Tests-Emergent Clinical Biomarkers.* Front Med (Lausanne), 2018. **5**: p. 248.
74. Caputo, R., et al., *Gene Expression Assay in the Management of Early Breast Cancer.* Curr Med Chem, 2020. **27**(17): p. 2826–2839.
75. Nikanjam, M., S. Kato, and R. Kurzrock, *Liquid biopsy: current technology and clinical applications.* J Hematol Oncol, 2022. **15**(1): p. 131.
76. Corces, M.R., et al., *The chromatin accessibility landscape of primary human cancers.* Science, 2018. **362**(6413).
77. Ren, B., et al., *High-resolution Hi-C maps highlight multiscale 3D epigenome reprogramming during pancreatic cancer metastasis.* J Hematol Oncol, 2021. **14**(1): p. 120.

78. Bi, W.L., et al., *Artificial intelligence in cancer imaging: Clinical challenges and applications.* CA Cancer J Clin, 2019. **69**(2): p. 127–157.
79. Kelly, T.K., D.D. De Carvalho, and P.A. Jones, *Epigenetic modifications as therapeutic targets.* Nat Biotechnol, 2010. **28**(10): p. 1069–78.
80. Rittiner, J., et al., *Therapeutic modulation of gene expression in the disease state: Treatment strategies and approaches for the development of next-generation of the epigenetic drugs.* Front Bioeng Biotechnol, 2022. **10**: p. 1035543.
81. Sarkar, S., et al., *Cancer development, progression, and therapy: an epigenetic overview.* Int J Mol Sci, 2013. **14**(10): p. 21087–113.
82. Costa, P., et al., *Epigenetic reprogramming in cancer: From diagnosis to treatment.* Front Cell Dev Biol, 2023. **11**: p. 1116805.

# Detection Methods for Epigenetic Mechanisms in Breast Cancer

7

Rachisan Djiake Tihagam and Sanchita Bhatnagar

**Abstract**

Epigenetic dysregulation of gene expression is central to breast cancer initiation, progression, and treatment response. These alterations include histone modifications, DNA methylation, and expression of noncoding RNA. Technological advancements have improved our ability to identify histone modifications, DNA methylation patterns and provided critical insights into the epigenetic regulation of gene expression and its role in breast oncogenesis. The epigenetic profiles of healthy and breast cancer tissues revealed several diagnostic and prognostic biomarkers. In this article, we review the methodologies commonly used to study tumor-associated histone modifications and DNA methylation changes in breast cancer, highlighting their principles, applications, and advancements, such as chromatin immunoprecipitation (ChIP), bisulfite conversion of DNA, next-generation sequencing (NGS), and chromatin immunoprecipitation sequencing (ChIP-seq).

R. D. Tihagam · S. Bhatnagar (✉)
Department of Medical Microbiology and Immunology, University of California Davis School of Medicine, Davis, CA, USA
e-mail: sbbhatnagar@ucdavis.edu

**Keywords**

Chromatin immunoprecipitation (ChIP) · Bisulfite conversion of DNA · Next generation sequencing (NGS) · Chromatin Immunoprecipitation Sequencing (ChIP-seq) · Bisulfite sequencing (Bis-seq) · Whole-genome bisulfite sequencing (WGBS) · Methylation-specific PCR (MSP)

## 7.1 Introduction

Breast cancer continues to pose a significant health concern as it is the most common cancer diagnosed as well as the second leading cause of cancer-related death among women worldwide [1]. Despite substantial advances in diagnosis and treatment, the exact mechanism underlying breast cancer heterogeneity remains unclear owing to its complexity. As a result, understanding the molecular mechanisms is critical to improving patient outcomes. In recent years, the emergence of novel technologies has revolutionized our ability to examine the complex molecular landscape of breast cancer, with a particular emphasis on the significance of epigenetic modifications such as histone modification and DNA methylation. These modifications control gene expression patterns and chromatin structure, regulating cellular phenotype and function. Histone modifications regulate chro-



S. Bhatnagar (ed.), *Epigenetic Mechanisms in Breast Cancer Therapy and Resistance*, Advances in Experimental Medicine and Biology 1465, https://doi.org/10.1007/978-3-031-66686-5_7

matin accessibility and gene expression. For example, the acetylation of histone H3 at lysine 9 (H3K9Ac) is associated with open and accessible chromatin regions, facilitating gene expression [2–5]. The ubiquitination of histone H2A at lysine 119 (H2Aub) decreases chromatin accessibility, resulting in gene silencing [6]. Likewise, histone methylation can either upregulate or downregulate the expression of genes depending on the specific amino acid residue and the methylation state [2–5]. For example, H3K4me1, H3K4me3, and H3K36me3 are associated with active transcription and are enriched at the open chromatin. H3K9me3 and H3K27me3 are associated with heterochromatin, indicating the transcriptionally silent states [1, 2, 5]. DNA methylation in gene promoter regions often leads to transcriptional repression of tumor-suppressor genes, contributing to cancer development and progression. For instance, silencing the tumor-suppressor CDKN2A in breast cancer is linked to hypermethylation of its promoter region [7].

Identifying the genome-wide profile of histone modifications and DNA methylation provides a comprehensive understanding of cell identity and disease states. Comparative analysis of epigenetic modifications reveals unique epigenetic signatures associated with normal and cancer cell cellular differentiation, developmental transitions, and pathological diseases such as breast cancer [4].

To assess global levels of histone modifications, numerous techniques such as Western blot (WB), immunohistochemistry (IHC), immunocytochemistry (ICC), and ELISA can be used, with antibodies targeting specific histone modifications. Moreover, chromatin immunoprecipitation (ChIP), hybridization to microarrays (ChIP-chip), and high-throughput sequencing (ChIP-seq) are powerful technologies that allow for not only the detection of histone modifications but also provide information about their genomic locations. These techniques have been extensively employed in fundamental research to investigate the functions of histone modifications in gene regulation, development, and disease, including cancer. Various established methods such as Bisulfite Sequencing (Bis-Seq), Whole-Genome Bisulfite Sequencing (WGBS), Methylation-Specific PCR (MSP), methylation arrays, or Methylation-Sensitive Restriction Enzyme Digestion can be employed to study DNA methylation patterns.

## 7.2 Histone Modification Profiling in Breast Cancer

ChIP is a powerful method to identify genomic regions bound by specific proteins, such as histone modifications or transcription factors. It allows researchers to map the binding sites of transcription factors, histone modifications, and other chromatin-associated proteins across the genome, providing valuable insights into gene regulatory networks and cellular processes. The ChIP assay involves several steps outlined below:

**Cross-Linking**
Histones are covalently cross-linked to their genomic DNA substrate in living cells using a common cross-linker agent like formaldehyde. This step is necessary to maintain the histone-DNA interactions.

**Cell Lysis**
Using ChIP lysis buffer, cells are lysed to release the chromatin, which consists of DNA wrapped around histone proteins, into solution.

**Chromatin Fragmentation**
Sonication breaks chromatin into smaller fragments, typically 200 to 600 base pairs in length. This step is crucial for efficient immunoprecipitation and peak resolution during ChIP-seq [8].

**Immunoprecipitation**
Antibodies against specific histone modifications are added to the sheared chromatin solution. Their specificity allows them to selectively bind to the histone-DNA complexes, capturing them and forming antibody-protein-DNA complexes. To help reduce background noise and increase the specificity of the immunoprecipitation, the antibody-histone-DNA complexes are washed to

remove nonspecifically bound histone and DNA fragments.

**DNA Precipitation/DNA Purification**

The bonds between the histone-DNA complexes are broken by reversing the cross-links by heat, which releases the DNA fragments associated with the histone of interest. The released DNA fragments are purified through phenol-chloroform extraction and ethanol precipitation. Accurate detection and quantification of histone modifications require proper sample preparation. Therefore, removing contaminants from DNA is crucial for the downstream analyses. The purified DNA fragments are subjected to library preparation, which involves fragment end repair, adaptor ligation, and emulsion PCR amplification. Subsequently, the libraries are loaded onto a DNA sequencer instrument for high-throughput sequencing.

ChIP-seq allows for genome-wide interrogation in a single sequencing run. In contrast, DNA microarrays require many arrays to cover the complete human genome, leading to high costs for comprehensive studies [9]. Consequently, ChIP-seq has largely replaced ChIP-chip for histone modification studies. Currently, there are four platforms available for the ChIP-seq experiment: 454 Genome Sequencer FLX developed by Roche, Sequencing by Oligo Ligation/Detection (SOLiD) developed by Applied Biosystems, true Single Molecular Sequencing (tSMS) developed by Helicos, and Genome Analyzer GA2 developed by Illumina [10]. These platforms have been utilized in ChIP-seq to investigate various aspects of epigenetic regulation, including histone modifications, transcription factor binding, and chromatin accessibility.

**454 Genome Sequencer**

This platform sequences longer reads and utilizes generic adapters annealed to beads. Its advantages include longer sequencing reads from 200 to 400 base pairs and lower sequencing bias. Its limitations include lower sequencing throughput and higher per-base sequencing costs than newer technologies like Illumina [11].

**Oligo Ligation/Detection (SOLiD)**

The SOLiD sequencing platform is based on sequencing by primer ligation, which incorporates the ligation of fluorescently labeled oligonucleotides into DNA fragments [12]. Although SOLiD is known for its high accuracy and robustness, it generates short sequencing reads, typically around 35 to 50 base pairs.

**True Single Molecular Sequencing (tSMS)**

Unlike the 454 Genome Sequencer FLX, GA2, and SOLiD sequencing methods, tSMS is not dependent on library building and PCR amplification but performs direct sequencing of single DNA molecules. Here, individual DNA molecules are immobilized on a solid surface and subjected to sequential cycles of nucleotide addition. Each nucleotide is fluorescently labeled, allowing for real-time detection of nucleotide incorporation [13]. tSMS reduces amplification bias, produces longer reads, and enables the sequencing of individual DNA molecules with increased accuracy and sensitivity. Its limitations include expensive costs and higher technical errors as even minimal DNA contamination can affect the results.

**Genome Analyzer GA2 from Illumina**

This platform utilizes sequencing-by-synthesis technology, where nucleotides are sequentially added to a growing DNA strand, and each incorporated nucleotide is detected through fluorescence, allowing for the determination of the sequence of each DNA fragment [14]. The GA2 platform is capable of sequencing millions of DNA fragments in parallel in a relatively short time, with lower per-base sequencing costs, high sequencing depth, and accuracy. However, the GA2 platform generates shorter sequencing reads and provides limited resolution for repetitive regions of genomes [14].

Each of these sequencing platforms has its strengths and limitations. Genome Analyzer GA2 from Illumina is the most widely used high-throughput sequencing technique for ChIP-seq studies.

## 7.3 Illumina Genome Analyzer GA2 ChIP-Seq Sequencing Workflow

1. *Library preparation:* Chip-seq libraries are prepared by isolating genomic DNA bound to specific histones using ChIP steps. The genomic DNA is fragmented into shorter fragments, and adapters are ligated at the 5′ and 3′ ends of the fragmented DNA. These short fragments are then PCR amplified and purified [15].
2. *Cluster generation*: The PCR-amplified ChIP-seq libraries are loaded onto the Genome Analyzer GA2 flow cell. The columns contain a high-density array of immobilized oligonucleotides complementary to the adapters, allowing short DNA fragments to be captured. Subsequently, through bridge amplification, clonal clusters are generated [8, 16].
3. *Sequencing:* The sequencing step involves repeatedly adding fluorescently labeled nucleotides to the clusters, which are then photographed to determine the sequence of each fragment and output raw base calls plus quality scores. This process is performed numerous times, generating short sequencing reads for each fragment [8].
4. *Image analysis and base-calling*: Illumina ChIP-seq data from the Genome Analyzer are first transitioned through the image analysis and base-calling phases. During sequencing, images of the flow cell are captured at each cycle to detect the fluorescent signals from the incorporated nucleotides. These images are analyzed to determine the sequence of each DNA fragment. The fluorescence signals are converted into nucleotide sequences using base-calling software. This step assigns a base (A, T, C, or G) to each position in the sequence based on the detected fluorescence signals.
5. *Alignment:* The short sequence reads are aligned to a reference genome using an alignment platform such as ELAND, Bowtie, HITSAT2, or STAR [17–19]. The alignment produces an output listing all the sequence reads with their respective genomic coordinates.
6. *Data visualization*: The sequencing reads data, which aligns uniquely to a reference genome, can be visualized as a customized track within the UCSC genome browser and the Integrative Genomics Viewer (IGV). This helps visualize ChIP-seq peaks, genomic annotations links with the enriched sequences, and other relevant data tracks in the genomic context.

## 7.4 DNA Methylation Profiling in Breast Cancer

Several techniques can be used to interrogate DNA methylation profiles in breast cancer tissues [20–22], including:

1. *Bisulfite Sequencing (Bis-Seq)* is the gold-standard approach for DNA methylation analysis. DNA is treated with sodium bisulfite to convert unmethylated cytosines to uracil. Following PCR amplification and sequencing, converted cytosines can be differentiated from methylated cytosines, allowing DNA methylation patterns to be determined at a single-base resolution.
2. *Whole-Genome Bisulfite Sequencing (WGBS):* WGBS is a comprehensive approach that includes bisulfite treatment of genomic DNA and then utilizes high-throughput genome sequencing. WGBS provides genome-wide coverage of DNA methylation patterns at single-nucleotide resolution, enabling the identification of methylated and unmethylated cytosines throughout the genome.
3. *Methylation-Specific PCR (MSP)* is a PCR-based method for selectively amplifying methylated or unmethylated DNA sequences using primers unique to the methylation status of the target region. MSP is frequently used in breast cancer samples to detect methylation alterations at specific loci, such as tumor-suppressor genes or oncogenes.
4. *Methylation Arrays:* Methylation arrays, such as Illumina's Infinium Methylation EPIC BeadChip or Methylation EPIC Array, enable high-throughput analysis of DNA methylation

patterns at CpG sites throughout the genome. These arrays are based on bisulfite conversion of DNA and probe hybridization to quantitate DNA methylation levels across CpG sites.

5. *Methylation-Sensitive Restriction Enzyme Digestion:* This procedure entails digesting genomic DNA with methylation-sensitive restriction enzymes, which break DNA at specific recognition sequences only if the cytosine is unmethylated. The resulting DNA fragments can be examined using PCR or sequencing to determine DNA methylation levels at restriction sites.

## 7.5 Conclusion

Histone modifications and DNA methylation analyses are critical in breast cancer. Several approaches for investigating these epigenetic changes are described above. These methods allow for the detection of dysregulated epigenetic modifications associated with subtypes of breast cancer, metastasis, and response to therapy. DNA methylation can be investigated using Bis-Seq, MS, and methylation arrays. These techniques provide information about the DNA methylation landscape of breast cancer genomes, such as identifying hypermethylated tumor-suppressor genes, hypomethylated oncogenes, and epigenetic biomarkers for early detection and prognosis. Overall, combining histone modification and DNA methylation profiling techniques in breast cancer research can improve our understanding of disease biology, identify novel biomarkers for diagnosis and prognosis, and guide the development of more effective targeted therapies for breast cancer patients.

## References

1. Siegel, R.L., A.N. Giaquinto, and A. Jemal, *Cancer statistics, 2024.* CA Cancer J Clin, 2024. **74**(1): p. 12–49.
2. Miller, J.L. and P.A. Grant, *The role of DNA methylation and histone modifications in transcriptional regulation in humans.* Subcell Biochem, 2013. **61**: p. 289–317.
3. Lawrence, M., S. Daujat, and R. Schneider, *Lateral Thinking: How Histone Modifications Regulate Gene Expression.* Trends Genet, 2016. **32**(1): p. 42–56.
4. Zhao, Z. and A. Shilatifard, *Epigenetic modifications of histones in cancer.* Genome Biol, 2019. **20**(1): p. 245.
5. Talbert, P.B. and S. Henikoff, *The Yin and Yang of Histone Marks in Transcription.* Annu Rev Genomics Hum Genet, 2021. **22**: p. 147–170.
6. Chen, J.J., D. Stermer, and J.C. Tanny, *Decoding histone ubiquitylation.* Front Cell Dev Biol, 2022. **10**: p. 968398.
7. Spitzwieser, M., et al., *Hypermethylation of CDKN2A exon 2 in tumor, tumor-adjacent and tumor-distant tissues from breast cancer patients.* BMC Cancer, 2017. **17**(1): p. 260.
8. Kidder, B.L., G. Hu, and K. Zhao, *ChIP-Seq: technical considerations for obtaining high-quality data.* Nat Immunol, 2011. **12**(10): p. 918–22.
9. Lee, T.I., S.E. Johnstone, and R.A. Young, *Chromatin immunoprecipitation and microarray-based analysis of protein location.* Nat Protoc, 2006. **1**(2): p. 729–48.
10. O'Geen, H., L. Echipare, and P.J. Farnham, *Using ChIP-seq technology to generate high-resolution profiles of histone modifications.* Methods Mol Biol, 2011. **791**: p. 265–86.
11. Patrick, K., *454 life sciences: illuminating the future of genome sequencing and personalized medicine.* Yale J Biol Med, 2007. **80**(4): p. 191–4.
12. Garrido-Cardenas, J.A., et al., *DNA Sequencing Sensors: An Overview.* Sensors (Basel), 2017. **17**(3).
13. Orlando, L., et al., *True single-molecule DNA sequencing of a pleistocene horse bone.* Genome Res, 2011. **21**(10): p. 1705–19.
14. Kircher, M., U. Stenzel, and J. Kelso, *Improved base calling for the Illumina Genome Analyzer using machine learning strategies.* Genome Biol, 2009. **10**(8): p. R83.
15. Sundaram, A.Y., et al., *A comparative study of ChIP-seq sequencing library preparation methods.* BMC Genomics, 2016. **17**(1): p. 816.
16. Sun, C.X., et al., *A Clustering Approach for Motif Discovery in ChIP-Seq Dataset.* Entropy (Basel), 2019. **21**(8).
17. Langmead, B. and S.L. Salzberg, *Fast gapped-read alignment with Bowtie 2.* Nat Methods, 2012. **9**(4): p. 357–9.
18. Li, H., et al., *The Sequence Alignment/Map format and SAMtools.* Bioinformatics, 2009. **25**(16): p. 2078–9.
19. Zhang, Y., et al., *Model-based analysis of ChIP-Seq (MACS).* Genome Biol, 2008. **9**(9): p. R137.
20. Kurdyukov, S. and M. Bullock, *DNA Methylation Analysis: Choosing the Right Method.* Biology (Basel), 2016. **5**(1).
21. Khodadadi, E., et al., *Current Advances in DNA Methylation Analysis Methods.* Biomed Res Int, 2021. **2021**: p. 8827516.
22. Harrison, A. and A. Parle-McDermott, *DNA methylation: a timeline of methods and applications.* Front Genet, 2011. **2**: p. 74.

# Index





S. Bhatnagar (ed.), *Epigenetic Mechanisms in Breast Cancer Therapy and Resistance*, Advances in Experimental Medicine and Biology 1465, https://doi.org/10.1007/978-3-031-66686-5

The manufacturer's authorised representative in the EU is Springer
Nature Customer Service Centre GmbH, Europaplatz 3, 69115 Heidelberg,
Germany. If you have any concerns regarding our products, please
contact ProductSafety@springernature.com

Printed and bound by CPI Group (UK) Ltd, Croydon, CR0 4YY
30/09/2026
02225607-0020